The Neuroscience of Handwriting

Applications for Forensic Document Examination

INTERNATIONAL FORENSIC SCIENCE AND INVESTIGATION SERIES

Series Editor: Max Houck

Scientific Examination of Documents: methods and techniques, 2nd edition
D Ellen
ISBN 9780748405800
1997

Forensic Examination of Human Hair
J Robertson
ISBN 9780748405671
1999

Forensic Examination of Fibres, 2nd edition
J Robertson and M Grieve
ISBN 9780748408160
1999

Forensic Examination of Glass and Paint: analysis and interpretation
B Caddy
ISBN 9780748405794
2001

Forensic Speaker Identification
P Rose
ISBN 9780415271827
2002

Bitemark Evidence
B J Dorion
ISBN 9780824754143
2004

The Practice of Crime Scene Investigation
J Horswell
ISBN 9780748406098
2004

Fire Investigation
N Nic Daéid
ISBN 9780415248914
2004

Fingerprints and Other Ridge Skin Impressions
C Champod, C J Lennard, P Margot, and M Stoilovic
ISBN 9780415271752
2004

Forensic Computer Crime Investigation
Thomas A Johnson
ISBN 9780824724351
2005

Analytical and Practical Aspects of Drug Testing in Hair
Pascal Kintz
ISBN 9780849364501
2006

Nonhuman DNA Typing: theory and casework applications
Heather M Coyle
ISBN 9780824725938
2007

Chemical Analysis of Firearms, Ammunition, and Gunshot Residue
James Smyth Wallace
ISBN 9781420069662
2008

Forensic Science in Wildlife Investigations
Adrian Linacre
ISBN 9780849304101
2009

Scientific Method: applications in failure investigation and forensic science
Randall K Noon
ISBN 9781420092806
2009

Forensic Epidemiology
Steven A Koehler and Peggy A Brown
ISBN 9781420063271
2009

Ethics and the Pracice of Forensic Science
Robin T Bowen
ISBN 9781420088939
2009

Introduction to Data Analysis with R for Forensic Scientists
James Michael Curran
ISBN: 9781420088267
2010

Forensic Investigation of Explosions, Second Edition
A Beveridge
ISBN 9781420087253
2011

Firearms, the Law, and Forensic Ballistics, Third Edition
Tom Warlow
ISBN 9781439818275
2011

The Neuroscience of Handwriting: Applications for Forensic Document Examination
Michael P. Caligiuri and Linton A. Mohammed
ISBN 9781439871409
2012

INTERNATIONAL FORENSIC SCIENCE AND INVESTIGATION SERIES

The Neuroscience of Handwriting

Applications for Forensic Document Examination

Michael P. Caligiuri, PhD
Linton A. Mohammed, MFS

CRC Press
Taylor & Francis Group
Boca Raton London New York

CRC Press is an imprint of the
Taylor & Francis Group, an **informa** business

Cover art produced using handwriting samples from a patient with significant handwriting impairment associated with Parkinson's disease. Color image of the human cortex compliments of Hauke Bartsch of The Brain Observatory, University of California, San Diego. The red region of this image corresponds to the left superior parietal lobe, the primary cortical region thought to govern handwriting.

CRC Press
Taylor & Francis Group
6000 Broken Sound Parkway NW, Suite 300
Boca Raton, FL 33487-2742

First issued in paperback 2021

ISBN 13: 978-0-367-77854-5 (pbk)
ISBN 13: 978-1-4398-7140-9 (hbk)

Library of Congress Cataloging-in-Publication Data

Caligiuri, Michael P.
The neuroscience of handwriting : applications for forensic document examination / Michael P. Caligiuri and Linton A. Mohammed.
p. cm. -- (International forensic science and investigation series ; 25)
Includes bibliographical references and index.
ISBN 978-1-4398-7140-9 (alk. paper)
1. Writing--Identification. 2. Signatures (Writing) 3. Legal documents--Identification. 4. Graphology. I. Mohammed, Linton A. II. Title.

HV8074.C25 2012
363.25'65--dc23 2011038465

Visit the Taylor & Francis Web site at
http://www.taylorandfrancis.com

and the CRC Press Web site at
http://www.crcpress.com

Contents

Series Preface—International Forensic Science Series

The modern forensic world is shrinking. Forensic colleagues are no longer just within a laboratory but across the world. E-mails come in from London, Ohio, and London, England. Forensic journal articles are read in Peoria, Illinois, and Pretoria, South Africa. Mass disasters bring forensic experts together from all over the world.

The modern forensic world is expanding. Forensic scientists travel around the world to attend international meetings. Students graduate from forensic science educational programs in record numbers. Forensic literature—articles, books, and reports—grows in size, complexity, and depth.

Forensic science is a unique mix of science, law, and management. It faces challenges like no other discipline. Legal decisions and new laws force forensic science to adapt methods, change protocols, and develop new sciences. The rigors of research and the vagaries of the nature of evidence create vexing problems with complex answers. Greater demand for forensic services pressures managers to do more with resources that are either inadequate or overwhelming. Forensic science is an exciting, multidisciplinary profession with a nearly unlimited set of challenges to be embraced. The profession is also global in scope—whether a forensic scientist works in Chicago or Shanghai, the same challenges are often encountered.

The International Forensic Science Series is intended to embrace those challenges through innovative books that provide reference, learning, and methods. If forensic science is to stand next to biology, chemistry, physics, geology, and the other natural sciences, its practitioners must be able to articulate the fundamental principles and theories of forensic science and not simply follow procedural steps in manuals. Each book broadens forensic knowledge while deepening our understanding of the application of that knowledge. It is an honor to be the editor of the Taylor & Francis International Forensic Science Series of books. I hope you find the series useful and informative.

Max M. Houck, PhD
Principal Analyst
Analytic Services, Inc.
Washington, DC

Foreword

As aptly noted in The National Academies of Science's 2009 Report, Strengthening Forensic Science in the United States: A Path Forward, "[t]he law's greatest dilemma in its heavy reliance on forensic evidence…concerns the question of whether—and to what extent—there is science in any given forensic science discipline." Your honor is indeed honored to write brief introductory remarks on this well-researched and well-written book which contains pertinent and reliable scientific knowledge integrated by the authors with respect to handwriting and signature authentication. These authors have taken the necessary steps to open doors to advancing the forensic science of handwriting forward by conducting their extensive neurobiological, neuroanatomical, and neurochemical research on how the complex regions of the brain such as the cortical and sub-cortical regions manage hand movements. They provide empirical data for the legal and scientific communities to understand how disease, medication, drugs, and the age process affect handwriting.

As a state trial judge of general jurisdiction for 22 years, I am impressed by the extensive work performed and contained within these fourteen chapters by these two well-qualified experts, Michael P. Caligiuri, Ph.D. and Linton A. Mohammed, MFS. Their book has three Parts: In Section I, the authors not only provide the backdrop for understanding motor control regarding handwriting but also describe how the aging process affects motor control and handwriting. In Section II, the authors explain the latest trends in the quantitative approach to signature authentication and how data revealing kinematic features of signatures provide pen pressures, stroke formations, and movement durations. These experts test hypotheses regarding "whether a signature is the product of highly programmed motor behavior (i.e., authentic) or a forgery (i.e., an attempt to 'overwrite' an internal handwriting program) to be tested in practice." Their work suggests to forensic document examiners that "accurate measures of stroke length and calculating the upstroke/downstroke ratio or difference can increase the scientific validity and reliability of judgments of authenticity." In Section III, the authors present their laboratorial data and conclusions regarding the effects that disease, medication, drugs, and the aging process have on handwriting.

With this book, these experts inspire us as scientific and legal professionals to further explore how disease, medication, drugs, and the age process affect handwriting. These two authors, through their extensive work, have begun the necessary dialogue for forensic document examiners, lawyers, judges, educators, and researchers, as recommended by the National Academies of Science, regarding the forensic sciences. These experts are not only "talking the talk," but are "walking the walk," by conducting empirical research with neurobiological, neuroanatomical, and neurochemical bases in order to validate whether and if so, how much science is within the field of signature and handwriting authentication.

Judge Stephanie Domitrovich, Ph.D.
Sixth Judicial District of PA

Preface

The neurobiological understanding of handwriting stems from decades of fundamental research in the fields of motor control, neuroscience, kinematics, and robotics. This book is an attempt to integrate these fields and facilitate a more scientific approach to the evaluation of questioned signatures and handwriting. This book comes at a time when the validity and reliability of document examination is being closely scrutinized. A review of the status of questioned document examination by the National Academy of Sciences in 2009 concluded that the scientific basis for the comparison of handwriting needed to be strengthened. The NAS report underscored the need for fundamental scientific inquiry into the validity and reliability of document examination. The *Daubert* trilogy[1] of judgments by the US Supreme Court has made it clear that scientific expert testimony must be based on science that is grounded in empirical research.

Decades of laboratory research in handwriting have given us the tools necessary to elucidate normal and pathological processes underlying handwriting and signature production. Unfortunately, these principles are rarely incorporated into modern research on forensic document examination. The overarching goal of this book is to educate the reader on the relevant neuroscientific principles underlying normal and pathological hand motor control and handwriting and to bridge the gap between theory and practice with examples from recent and ongoing laboratory studies.

The idea for this book grew from discussions during and following two workshops presented to the annual American Academy of Forensic Sciences and the American Society of Questioned Document Examiners (ASQDE) meetings held in 2010 entitled "Signature Examination Translating Basic Science to Practice." While these workshops explored a wide range of topics, including the neuroanatomy and neurochemistry of motor control, disease conditions, and medication and drugs that affect handwriting, and kinematic approaches to quantifying these effects, the workshop format allowed for only surface treatment of these important topics. The many intuitive questions, case presentations, and thoughtful discussions that took place during these workshops were a valuable impetus for the organization and content of this book.

The book is organized into three main parts. In Section I, we provide a general background on the fundamentals of motor control, with specific reference to handwriting. Fundamental principles in the neuroanatomy and neurochemistry of hand motor control are presented in Chapter 1. Chapters 2 and 3 provide backgrounds in theories of motor control and their application to research in handwriting, respectively. Chapter 4 presents an overview of common neurodegenerative diseases such as Parkinson's disease, Huntington's disease, multiple sclerosis, motor neuron disease, Alzheimer's disease, essential tremor, and others. This chapter focuses on the epidemiology, pathophysiology, and motor characteristics of neurological disease. In Chapter 5, we review common psychotropic medications prescribed for depression, bipolar disorder, and psychosis; their mechanisms of action; and why they are important in understanding motor behavior and handwriting. Section I concludes with an overview of the aging process and its effects on motor control and handwriting in Chapter 6.

Section II includes three chapters on advances made in the quantitative approach to signature authentication. Chapter 7 begins with an extensive overview of the kinematic approach and describes new findings on the kinematic analyses of genuine, disguised, and forged signatures. While the vast majority of research regarding signatures has focused on static traces, modern technology has enabled researchers to quantify the kinematic features of signatures at the level of an individual pen stroke. Historically, visually detectable features in handwritten signatures formed the basis of evidence supporting whether a questioned signature was genuine, disguised, or forged (Michel 1978; Herkt 1986; Mohammed 1993; Wendt 2000). Today, research into static features associated with different signing behaviors can be supplemented by dynamic studies where kinematic data are collected from subjects' signing on digitizing tablets. This technique has been used to report on the effects of disguise and simulation behaviors in terms of pen pressure, stroke formation, and movement duration (e.g., van Gemmert et al. 1996).

Data from the authors' laboratories are presented in Chapters 8 and 9. These chapters review the literature and present current laboratory research further bridging the gap between theory and practice. Based on our understanding of the principles of motor control, we are able to test specific hypotheses about whether a signature is the product of highly programmed motor behavior (i.e., authentic) or a forgery (i.e., an attempt to "overwrite" an internal handwriting program) to be tested in practice.

The effects of disease, medication, and aging pose additional challenges to the forensic document examiner, as these effects tend to increase the range of variation of a writer's signature and reduce certainty. The wider the range of variation is, the more difficult it becomes to identify characteristics of a contemporary genuine signature. The majority of studies reported in the document examination literature comprise case studies rather than empirical research. Hilton (1969) reported that in cases involving writers in poor health, "expert decisions in this class of case are far from simple" and further noted that "signatures executed during illness or advanced age may be very erratic and poorly written." In discussing the identification of signatures and diagnosing mental illness from handwriting, Hilton notes that while identification is possible by the forensic document examiner (FDE), attempts at diagnosis lead to mediocre results (Hilton 1962).

Unfortunately, with the exception of a few dozen pages in the book by Huber and Headrick (1999) and a handful of peer-reviewed articles, researchers have not utilized modern scientific methods to further the understanding of the effects of medication and disease on handwriting. To fill this gap, Section III presents current results from our laboratory on these important influences on handwriting. Chapter 10 extends the fundamental principles of neurological diseases and their effects on motor control (Chapter 4) to the laboratory, where systematic research on the effects of these influences on handwriting are presented. In Chapter 10 we present findings from prior and ongoing research from our laboratory on handwriting in Parkinson's disease, essential tremor, progressive supranuclear palsy, and Alzheimer's disease. Chapters 11 and 12 focus largely on the effects of psychotropic medications and substance abuse on handwriting, respectively. Chapter 13 concludes this section with a summary of empirical research on the effects of aging on handwriting.

We hope the book will have wide appeal to the forensic document examiner community, the legal community, and educators and researchers in the fields of motor control and clinical neuroscience. For those seeking to understand the interactions between variability in the brain's response to disease and medications taken to treat disease and the extraordinary and complex process of handwriting, we hope this book raises new questions and opens new doors to the scientific process of signature and handwriting authentication.

MPC
LAM

Note

1. The *Daubert* trilogy refers to the three US Supreme Court cases that articulated the *Daubert* standard: *Daubert v. Merrell Dow Pharmaceuticals*, which held that Rule 702 did not incorporate the *Frye* "general acceptance" test as a basis for assessing the admissibility of scientific expert testimony; *General Electric Co. v. Joiner*, which held that an abuse-of-discretion standard of review was the proper standard for appellate courts to use in reviewing a trial court's decision of whether expert testimony should be admitted; and *Kumho Tire Co. v. Carmichael*, which held that the judge's gatekeeping function identified in *Daubert* applies to all expert testimony, including that which is nonscientific.

Acknowledgments

Several individuals played key roles in the development of our thinking and inspiration for this book. We acknowledge our academic mentors and colleagues for their unending interest in and intellectual support of our research and pursuit for greater understanding of human movement and its pathologies—spanning, for some individuals, over 25 years. Their collective knowledge and wealth of experience made this book possible. Particularly, we acknowledge Drs. James Lohr and Hans-Leo Teulings for many years of thoughtful discussion, designing studies, and evaluating data, much of which is included in this book, and Drs. Douglas Rogers and Bryan Found for their guidance in helping us bridge the gap between theories of handwriting motor control and forensic applications. Much of the research conducted in our laboratory would not have been possible without support from the National Institutes of Mental Health and the National Institute of Drug Abuse. We extend sincere appreciation to the volunteers and patients who participated in our research. Lastly, and most importantly, we thank our wives, Debra and Fanzia, for their support throughout this process and for listening to us ramble on about pen stokes, kinematics, and neural circuits, so that eventually it began to make sense, at least for now.

The Authors

Michael P. Caligiuri, PhD, is a professor in the Department of Psychiatry at the University of California, San Diego. His research over the past two decades has focused on understanding how drugs and disease affect motor control and fine hand movements. He has served as the lead scientist on several federally and industry-sponsored studies on identifying treatment response in psychiatric patients and has authored over 100 peer-reviewed articles in medical journals and book chapters on movement disorders, brain imaging, and biomedical instrument development. His current research interest focuses on kinematic studies of impaired handwriting and understanding writer-based sources of variability in signature authentication.

Linton A. Mohammed, MFS, D-ABFDE, has been a forensic document examiner for 25 years. He has testified as an expert witness over 100 times in the United States, England, and the Caribbean. He is currently in private practice with Rile, Hicks, & Mohammed with offices in Long Beach and San Bruno, California. He is certified by the American Board of Forensic Document Examiners and holds a diploma in document examination from the Forensic Science Society in England. He is the current president of the American Society of Questioned Document Examiners, is a fellow in the Questioned Document Section of the American Academy of Forensic Sciences, and is currently completing work for his PhD in human biosciences at LaTrobe University, Melbourne, Australia.

The Authors

I

Fundamental Aspects of Motor Control and Handwriting

Neuroanatomical and Neurochemical Bases of Motor Control

1

Introduction

The human brain is a complex system governing automatic and willed behaviors, multimodal perception, emotion, and restorative functions. It is without doubt that brain function today is the refinement of millions of years of adaptation and evolution. With few exceptions, the nervous system control functions we observe in humans today can be traced to corollary functions of lower animals. For example, there is considerable evidence demonstrating a relationship between cranial capacity and hand morphology and function over the past 1.75 million years. Fine motor control of the hand and articulatory system for speech are perhaps the most obvious among the many evolutionary advances that can be traced to an increase in brain size and complexity.

Historical Perspective on Brain Function for Hand Motor Control

Numerous writings can be found in the literature on brain function throughout antiquity. Much of this literature is nicely summarized in a very readable treatise by Stanley Finger (1994). The idea that different parts of the brain subserved different functions may be traced to the writings of the Roman physician Galen (AD 130–200). Galen's anatomical work with various animals showed that the cerebrum was softer than the cerebellum, leading to his conclusion that motor and sensory pathways were separate. He further reasoned that, unlike the motor nerves, sensory nerves needed to be pliable to retain the sensory information for long periods of time. Galen thus asserted that the sensory nerves went to the cerebrum while the motor nerves went to the cerebellum because the former was softer than the latter.

Ventricular localizationalists dominated brain science throughout the fourth and fifth centuries. Figure 1.1 depicts the neuroanatomical understanding of brain localization of the 1200s as envisioned by Albertus Magnus (1206–1280). Throughout the years following Galen, the dominant theory held that higher brain functions such as cognition, imagination, and memory

Figure 1.1 Drawing of the ventricles by Albertus Magnus published in the 1506 edition of *Philosophia naturalis.* (Photo source: Corbis, with permission.)

were associated with the cerebral ventricles. The ventricles were where the spirits from the sensory nerves ended before being taken up by the motor nerves to invoke action.

Functional localization in the ventricles was widely accepted for hundreds of years, even into the Middle Ages. For example, in 1481, the Italian physician Antonio Guainerio described two patients: one who was unable to speak more than a few words at a time (a condition we now refer to as aphasia) and another who could not remember people's names. Assuming both conditions stemmed from a memory disorder, Guainerio diagnosed their problems as stemming from excessive buildup of phlegm in the posterior ventricle. It was not until the early 1500s that the ventricular doctrine began to unravel. During this time, Leonardo da Vinci dissected hundreds of brains from cadavers and conducted experiments on ventricles from cattle brains. His observations were largely inconsistent with the assertions held by the ventricular localizationists of the time. While reasoning that the flow of "nervous spirit" from sensory nerves should be more midline than lateral, daVinci fell short of openly challenging the doctrine that higher mental functions were seated in the ventricles.

Others, however, argued for a completely different view of the cerebral functional localization and the role of the ventricles. Andreas Vesalius (1514–1564) rekindled interest in brain function during the Renaissance. At age 23, Vesalius received a grant (in the form of material) from the Senate of the Republic of Venice to conduct public dissections. Books on his work published in 1543 set the stage for a dramatic paradigm shift in the structure–function relationship of the human brain. Vesalius's main argument was that since the human ventricular system was not different in shape from other mammals and since other animals lacked higher reasoning powers, how could these powers be relegated to the ventricles? In describing Galen's work, he uncovered nearly 200 cases in which Galen's anatomic drawings were incorrect. While generally opposed to the idea of ventricular localization, Vesalius did not reject the traditional view that animal spirits were produced in the ventricles. His progressive stance on anatomy was dissociated from his adherence to traditional principles of physiology. More than 1,300 years after Galen, Vesalius wrote that the ventricles are no more than spaces into which air flows to be mixed with vital spirit from the heart and then transformed into animal spirit distributed through the nerves to organs of sensation and motion.

A century after Vesalius's death, Thomas Willis (1621–1675) published a book entitled *Cerebri Anatome* in which he proposed that the cerebral gyri controlled higher cognitive functions. Vital and involuntary functions were attributed to the cerebellum (along with what we now refer to as the midbrain and pons). The corpus striatum was thought to play a role in sensation and movement. Willis had effectively launched the post-Renaissance idea that individual brain parts contributed to different functions.

The first truly accurate theory of cerebral localization appeared in the mid-1700s. Emanuel Swedenborg (1688–1772) postulated that different functions were represented in different anatomical loci within the cerebral cortex. He argued that the variation in clinical signs observed from individuals with brain trauma could only be explained by anatomical separation of function. He identified distinct cerebral regions separated by fissures and gyri. He placed the motor cortex in the anterior portion of the brain and further identified a somatotopic representation by which the muscles of the extremities were controlled by upper convolutions, the trunk by the middle convolutions, and the neck and head by the lower convolutions. Unfortunately for him, his work was not widely distributed until after his death. While Swedenborg was developing his ideas of cerebral localization, he "began to experience mystical visions" (Finger 1994, p. 30), which led him to abandon his work in the neurological sciences in favor of a religious following.

The modern era of cerebral localization and functional specificity began in the 1800s with the writings of Bell (1774–1842), Bouillaud (1796–1881), Andral (1797–1846), and Broca (1824–1880). Much of the early work was in reaction against phrenology, the concept promoted by Franz Joseph Gall and his devoted follower Johann Spurzheim (1776–1832). Phrenology is the pseudoscience of attributing brain function with structure primarily by examining the surface of the cranium. Franz Joseph Gall collected over 300 skulls of individuals from the extremes of society (scholars, statesmen, criminals, lunatics) and attempted to correlate mental characteristics with cranial surface maps. Phrenology was adopted by physicians of the time for diagnosing neurological disease, by criminologists for attributing criminal behavior to a physically defective brain, and by scholars for selecting individuals with particular intellectual or artistic talents.

Phrenology remained popular throughout Europe and the Unites States from the 1780s until Gall died in 1828. Debate dominated the scientific community throughout the 1800s with conservatives following the lead of Marie-Jean Flourens (1774–1867), who advocated greater emphasis on laboratory and animal study, while liberal followers of Aubertin (1825–1893) and others advocated localization based on clinical evidence. Conservatives cautioned against the direct structure–function theory on the basis of inconsistent laboratory studies. Localizationalists, on the other hand, advanced series after series of clinical cases supporting specific functions (e.g., speech or memory) to autopsy confirmed cortical regions. The two camps merged with experimental confirmation of Broca's clinical report of functional localization of motor behavior.

In 1869, Eduard Hitzig (1838–1907) and Gustav Fritsch (1838–1927) conducted an experiment (on a dog) proving that cortical localization need not be limited to a single function. Hitzig and Fritsch's experiment confirmed that applying electrical current to the frontal cortex in close proximity to the Broca's motor speech area impaired motor function. Hitzig and Fritsch replicated their finding in other animals and found distinctive cortical sites that elicited motor responses throughout the extremities, neck, and head on the opposite side of the stimulation. Further mapping led to unequivocal support for the existence of the motor cortex. Sir David Ferrier (1843–1928) replicated Hitzig and Fritsch's work and extended it to the monkey brain. Using more precise electrical stimulation and careful mapping, Ferrier was able to map a region of the motor cortex that corresponded to movement of a single finger. Ferrier's work was summarized in an 1876 publication entitled *The Functions of the Brain* that led neurosurgeons at the time to rely on functional maps for guidance during surgery.

Perhaps the most successful example of cortical mapping from the 1800s was the numbering system published by Korbinian Brodmann in 1909. His map of 52 discrete cortical regions was based on differences in structural and cellular composition. The map clearly distinguished motor from sensory areas. The map accurately delineated regions with fine granularity despite variation in experimental methods and across species. The histological delineation of cortical areas inevitably corresponded to functional specificity. Today, modern neurosurgery relies on many of the same principles of cortical mapping pioneered by the localizationalists throughout the 1800s.

In the late 1800s, Sigmund Exner (1846–1926) published a book entitled *Untersuchungen über die Lokalisation der Functionen in der Grosshirnrinde des Menschen* (*Studies on the Localization of Functions in the Cerebral Cortex of Humans*). In his book, Exner (1881) described a specific area of the cerebral cortex (the posterior part of the left middle frontal gyrus), which he attributed to handwriting. This area rapidly became known as Exner's "writing center." Based on just a few cases, Exner claimed that the posterior portion left middle frontal gyrus was the writing equivalent of Broca's motor speech area.

Despite having sparse data to back up this claim, the notion of a specific writing center ignited passionate debate throughout the scientific community (Roux et al. 2010). In his book, Exner described only four cases with agraphia that he associated with lesions to this area. Unfortunately, closer inspection revealed that in only one of these cases was the agraphia not accompanied by either hemiparesis or aphasia, which would lead to writing difficulties for reasons other than execution of the handwriting motor program (e.g., muscle weakness or paralysis or expressive language impairment).

As we will see later in this chapter, modern science has failed to support the notion of a single localized writing center. Rather, this "writing center" is likely to involve a network of cortical areas. Nonetheless, modern writers continue to refer to Exner's area as a writing center (e.g., Seitz et al. 1997; Sugihara, Kaminaga, and Sugishita 2006). However, because of its role in language processing, the posterior portion of the left middle frontal gyrus is not likely to be a member of this putative network.

Neuroanatomical Bases of Hand Motor Control

Functional Organization

Prior to undertaking a discussion of the anatomical regions, pathways, and circuits underlying hand motor control, it is important to understand the general organizational structure of the human nervous system. By understanding the fundamental organization of the brain, we can formulate hypotheses or predictions about what to expect in the form of altered handwriting following localized injury to the brain. In this section, we will introduce several approaches to understanding how the brain is organized for hand movement. Most of the organizational schemes hold that motor functions can be either spatially or topographically mapped onto a given brain region or network. Other organizational schemes rely on a hierarchical approach whereby certain brain areas (e.g., cortex) exhibit high-level, integrative or executive functions while other areas (e.g., basal ganglia) function in fine-tuning.

From the perspective of gross neuroanatomy, the nervous system can be divided into the central nervous system (CNS), consisting of the brain and spinal cord, and the peripheral nervous system (PNS), consisting of the nerves running to and from the spinal cord and periphery (i.e., muscle). Within the CNS, brain functions may be further organized into anterior–posterior or left–right dimensions. At this level, motor functions are typically attributed to anterior regions, while sensory processes are attributed to posterior regions. Cortical representation of the musculoskeletal system is bilateral. That is, sensory-motor functions of the left side of the body are regulated, at least at the cortical level, by the contralateral or right cerebral cortex, whereas sensory-motor functions of the right side of the body are regulated by the left cerebral cortex. This lateralization is well preserved throughout the cortex and spinal cord. Figure 1.2 shows the four main lobes of the left cerebral cortex including the frontal, parietal, temporal, and occipital cortices. Demarcation boundaries are based on Brodmann's cytoarchitectonic maps (see Figure 1.4 later in the chapter)

Another approach to understanding the functional organization of the CNS is based on the principle that the representation throughout the CNS is topographically organized. That is, the brain's functional organization for a given body area (e.g., the hand) is represented by a spatial map that is preserved throughout the brain's vertical hierarchy from the cortex through the basal ganglia and the brain stem and into the spinal column. This topographic representation generally holds that representation of lower extremities is topographically represented toward midline regions of the brain and, as we move from midline to lateral regions of the brain, representation follows from upper extremities to head and face, respectively. This scheme is

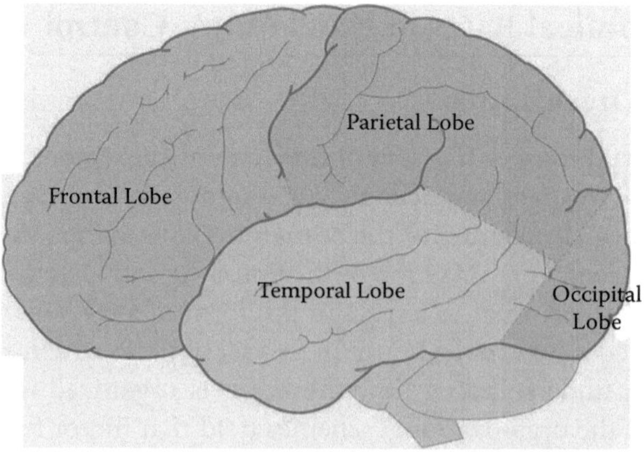

Figure 1.2 Demarcation boundaries are based on Brodmann's cytoarchitectonic maps.

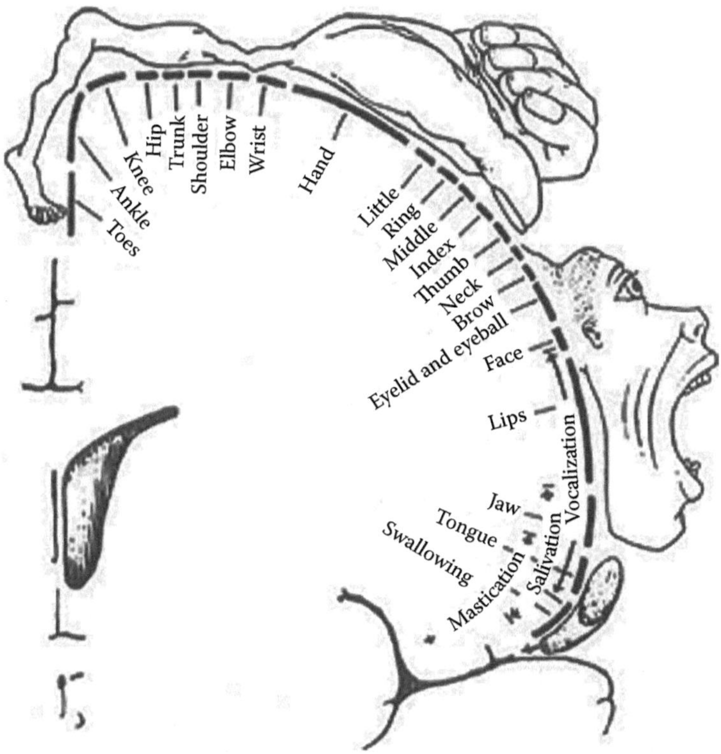

Figure 1.3 Drawing of the motor homunculus. The homunculus depicts a model of the human brain to illustrate anatomical representation of body movement. (From Penfield, W., and Rasmussen, T. 1950. *The Cerebral Cortex of Man: A Clinical Study of Localization of Function.* New York: MacMillan. Copyright Gale, a part of Cengage Learning, Inc. Reproduced by permission. www.cengage.com/permissions.)

commonly portrayed as a homunculus, or a representation of an anatomically deformed human being mapped onto the brain surface. Such a homunculus is depicted in Figure 1.3.

The precise control of hand movement for handwriting involves multiple motor and sensory areas throughout the central and peripheral nervous systems. In the following sections, we review the available evidence for specific roles of the motor and association cortices, basal ganglia, cerebellum, brain stem, and spinal cord in the control of handwriting.

Motor and Association Cortices

Three cortical regions play key roles in hand motor control: the primary motor area (Brodmann area 4), the premotor area (Brodmann area 6), and the supplementary motor area (SMA; located midline to area 5). Figure 1.4

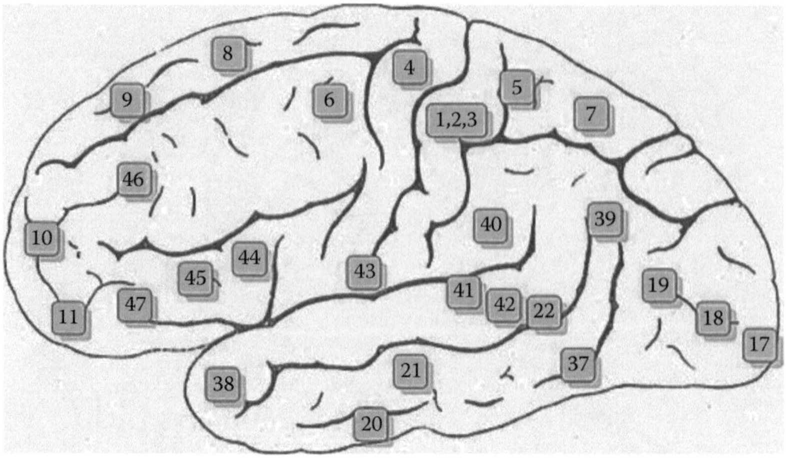

Figure 1.4 Broadmann's map of distinct cortical areas.

shows the demarcations of these areas in a diagram of the cortex. While the primary motor area and the premotor area extend from the medial to lateral portions of the cerebral hemisphere, the SMA extends into the mesial surface of the cerebral hemisphere within the longitudinal fissure that separates the two hemispheres (not shown in the diagram).

The principal function of the primary motor cortex in hand motor control is control over fine movement of distal musculature such as finger movement. Area 4 has a key role in the selection and sequencing of muscle contractions. As will be described later, this region of the cortex receives input from the somatosensory system for the regulation of appropriate grip and pressure necessary for handwriting. The premotor cortex is thought to play a key role in the sensory guidance of purposeful movement, particularly visual guidance. This area is involved in the coordination of activity from different muscle groups.

Area 6 receives input from the cerebellum to facilitate control of the duration of muscle firing (necessary to regulate movement distance) and the sequencing of muscle firing (necessary to regulate timing). The SMA is the primary target of projections from multiple areas of the brain involved in complex movement. It is thought that the SMA has a "clearinghouse" role as input from subcortical and sensory brain areas (see later discussion) are integrated within the SMA for delivery to the primary motor area. It is in this capacity that the SMA is considered important for the development and execution of motor programs.

Compelling evidence from studies of electrical stimulation in laboratory animals supports unique properties for areas 4 and 6 and the SMA in motor behavior (Eyzaguirre and Fidone 1975). Low-threshold electrical stimulation to each of these areas yields reproducible motor responses in the contralateral limb. Stimulation to areas 4 and 6 produces localized and somatotopically organized responses. However, electrical stimulation to the SMA produces movements that are less localized and more complex and that require higher thresholds than for areas 4 and 6. Moreover, electrical stimulation to the SMA usually leads to a bilateral motor response. This suggests a more complex integrative role of the SMA in motor behavior compared to the precentral motor areas.

Basal Ganglia and Extrapyramidal System

Just beneath the cerebral cortices reside several bodies of gray matter known collectively as the basal ganglia. They include the striatum (composed of the caudate and putamen) and the globus pallidus (comprising internal [GPi] and external [GPe] segments). Several other nuclei located in close proximity to the basal ganglia are equally important in motor control. These include the subthalamic nucleus (STN), the substantia nigra pars compacta (SN), and thalamus. For the purpose of this chapter, we consider this collective region of subcortical nuclei to constitute the extrapyramidal system, a term used when referring to brain regions involved in motor function outside the descending cortical-pyramidal (i.e., the cortex and brain stem) pathways. Figure 1.5 shows a coronal section of an MRI scan of a normal human brain.

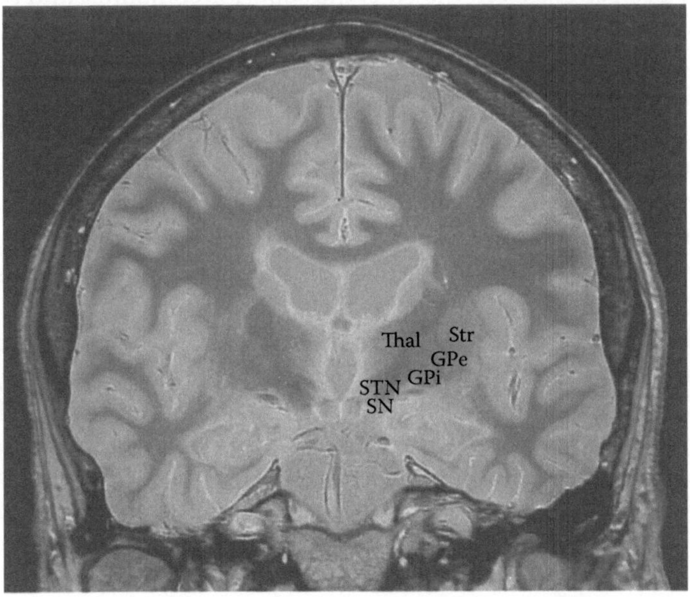

Figure 1.5 MRI image of a coronal section of the brain showing location of subcortical nuclei involved in motor function. Shown are locations of the striatum (Str), external (GPe) and internal (GPi) segments of the globus pallidus, the subthalamic nucleus (STN), the substantia nigra (SN), and the thalamus (Thal). (Image courtesy of The Brain Observatory, UCSD.)

The image shows the locations of the striatum (Str), the two segments of the globus pallidus (GPe and GPi), the subthalamic nucleus (STN), the substantia nigra (SN: pars compacta), and the thalamus (Thal). With the exception of the subthalamic nucleus, the primary function of the basal ganglia is to inhibit neurotransmission from cortical and subcortical centers.

The basal ganglia and neighboring subcortical nuclei play a key role in the maintenance and stabilization of voluntary movements, regulation of muscle tone, and integration of afferent information from the periphery. These nuclei receive input from all cortical areas and project to premotor and frontal cortices and the brain stem.

In a simple model, a motor command (e.g., to move a finger) originates in the premotor area and is forwarded to the primary motor area for selection of muscles and generation of muscle force. The command to contract a muscle is then transferred to parallel descending pathways: One projects neuronal excitation to the striatum and another to lower motoneurons terminating in the brain stem and spinal cord. The striatum, in turn, feeds neuronal excitation to the globus pallidus.

The striatum receives both excitatory and inhibitory inputs. The entire cortex, thalamus, amygdala, and hippocampus send excitatory projects to the striatum, while the GPe sends inhibitory projections to other subcortical regions and the thalamus. The primary inputs to the subthalamic nucleus are inhibitory projections from the GPe, the superior colliculus, and the cortex.

The primary output nucleus of the basal ganglia is the striatum. Numerous and complex circuits from the striatum project highly processed output throughout the brain to complete an important regulatory feedback loop. These circuits involve direct and indirect inhibitory projections from the striatum to brain stem, subcortical, and cortical centers. The direct pathway projects to the substantia nigra pars compacta, an important mechanism for modulating dopamine output. The indirect pathway consists of output from the ventral striatum to the substantia nigra, globus pallidus, subthalamic nucleus, thalamus, and pedunculopontine nucleus.

The direct pathway extends from the striatum to the GPi, which in turn projects to the thalamus and then cortex. The indirect pathway extends from the striatum to the subthalamic nucleus and then to the GPe. Fibers from the GPe also project to the GPi, thus forming a complete loop from cortex to striatum, to pallidum, to thalamus, and back to cortex. Figure 1.6 shows a block diagram representing this cortico–striato–pallido–thalamic (CSPT) loop.

The direct and indirect striatopallidal pathways are critical for control of fine movements. Functional or structural damage to either of these pathways can result in a movement disorder involving handwriting. When the direct pathway is activated, inhibitory pathways suppress tonically active neurons in the GPi. Because the globus pallidus sends inhibitory projections

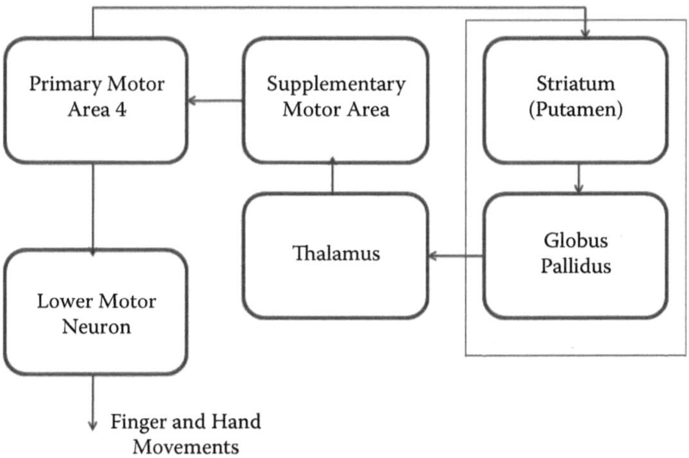

Figure 1.6 Block diagram representing the cortico–striato–pallido–thalamic (CSPT) circuit.

to the thalamus, the direct pathway serves to modulate pallidothalamic tone. Increased pallidothalamic tone activates thalamocortical excitation; decreased pallidothalamic tone suppresses thalamic activity to the cortex.

The indirect pathway has an additional waypoint. The indirect pathway comprises an inhibitory projection from the striatum to the subthalamic nucleus that, in turn, sends an excitatory projection to the GPe. This additional relay provides negative feedback within the striato–pallido–thalamic circuit. The direct and indirect pathways have opposite effects. Thus, the direct striatopallidal pathway *suppresses* pallidothalamic inhibition (and increases thalamocortical excitation), while the indirect pathway *increases* pallidothalamic inhibition (and decreases thalamocortical excitation). The direct pathway facilitates movement, while the indirect pathway suppresses movement. Figure 1.7 highlights the direct and indirect striatopallidal pathways.

Basal Ganglia Neurochemistry

The transmission of information throughout the basal ganglia and their communication with cortical and brain stem areas rely heavily upon neurotransmitters and neuromodulators. Neurotransmitters are endogenous chemicals that are released at nerve junctions and allow electrical impulses to pass from one neuron to another. Neurotransmitters can increase the likelihood that the electrical impulse will reach a critical threshold, thereby maintaining the flow of electrical impulses from one nerve ending to another (i.e., excit-

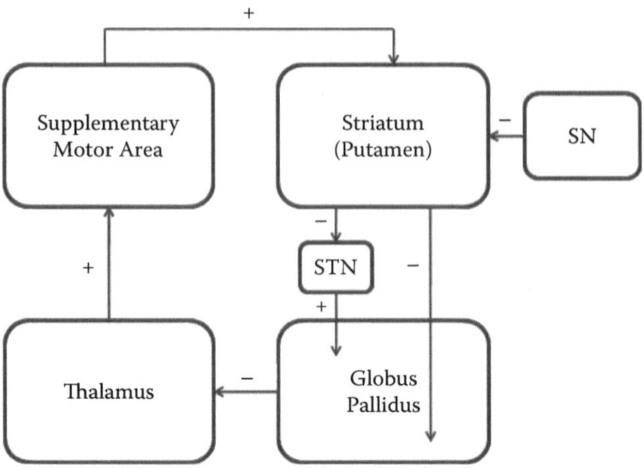

Figure 1.7 Block diagram of the CSPT circuit showing the direct (striatopallidal) and indirect (striato–subthalamic–pallidal) pathways. Positive signs indicate excitatory pathways; negative signs indicate inhibitory pathways.

atory), or decrease the likelihood that this threshold will be attained, thereby suppressing the flow of electrical impulses (i.e., inhibitory).

The primary excitatory neurotransmitter is glutamate; the primary inhibitory neurotransmitter found throughout the basal ganglia is γ-aminobutyric acid (or GABA). Dopamine serves to modulate both glutamatergic and GABAergic transmission by selectively inhibiting specific output nuclei within the basal ganglia. Unlike glutamatergic or GABAergic receptors (the molecular binding sites that reside on the nerve terminals), dopamine receptors can be either excitatory (D1) or inhibitory (D2). The excitatory D1 receptors are found on nerve terminals that are part of the direct striatopallidal pathway, whereas the inhibitory D2 receptors are found on nerve terminals that project within the indirect pathway. Both function to decrease thalamocortical inhibition (see previous discussion) and thus facilitate movement. Dopaminergic projections to the striatum can increase or decrease GABAergic inhibition to the globus pallidus. In this sense, dopamine is considered a neuromodulator. Other neurotransmitters located throughout the basal ganglia that subserve motor functions include acetylcholine, serotonin, and norepinepherine. Table 1.1 summarizes the role of key neurotransmitters and modulators within the basal ganglia.

The diagram in Figure 1.8 shows an overlay of the neurotransmitters onto the basic CSPT circuit. In this scheme, positive signs indicate excitatory neurotransmission involving glutamate, while negative signs indicate inhibitory neurotransmission involving GABA and dopamine (DA). Depending on the nature of the motor command, the globus pallidus may either excite or inhibit

Table 1.1 Neurotransmitters and Neuromodulators Found within the Basal Ganglia and Subcortical Brain Regions and Their Locations and Role in Motor Control

Neurotransmitter	Location	Activity	Function
Acetylcholine	Striatum	Excitatory and inhibitory	Sets tone for striatopallidal control of movement
Serotonin	Raphe nucleus	Modulatory	Mood regulation
Dopamine	Substantia nigra	Modulatory	Sets tone for striatopallidal control of movement
GABA	Globus pallidus, striatum	Inhibitory	Regulates basal ganglia excitability
Glutamate	Cortex, subthalamic nucleus, thalamus	Excitatory	Drives corticobasal ganglia and thalamocortical excitation

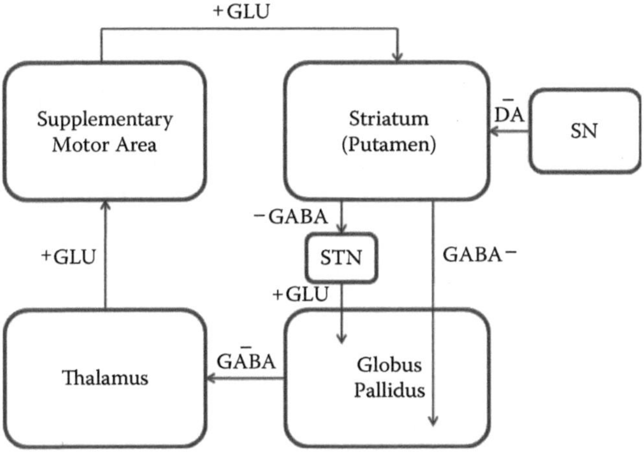

Figure 1.8 Block diagram of the cortical and subcortical areas completing the CSPT circuit and their excitatory (+) and inhibitory (–) neurotransmitters and neuromodulators. GLU = glutamate; DA = dopamine; GABA = γ-aminobutyric acid

neuronal firing of the thalamus. If the command calls for increased muscle force, for example, the globus pallidus would fire in such a way to suppress the inhibitory neurotransmitter GABA, enabling an increase in thalamic excitation. If, on the other hand, the command calls for a decrease in muscle force, the globus pallidus would fire in such a way to facilitate the inhibitory neurotransmitter GABA, leading to a decrease in thalamic excitation.

The thalamus, acting as a relay station, feeds the net neuronal firing state to multiple areas throughout the brain, particularly the SMA. Thus, the SMA is the recipient of the highly processed neuronal firing patterns that originated in the premotor cortex and is modulated by the basal ganglia before being forwarded to the thalamus and onto the motor cortex. The SMA acts as a comparator or buffer between the intended motor command and the observed movement by integrating afferent information (in the form of neuronal firing patterns) from the basal ganglia and sensory feedback from the periphery via thalamic projections.

SMA projections to the primary motor area are constantly updated and refined based on ongoing "calculations." It is through this network of descending (cortex to basal ganglia) and ascending (basal ganglia to thalamus to cortex) projections that the extrapyramidal system is intimately involved in fine-tuning complex movements, such as handwriting.

The preceding scenario describing the interaction between motor cortical and subcortical areas in generating, modulating, and executing motor commands is an oversimplification. The primary functions of the basal ganglia, STN, SNc, and thalamus are to regulate neuronal excitability and to ensure that complex movements are executed with the desired timing, precision, and force. This is accomplished through a complex network of excitatory and inhibitory pathways. One such network involving the globus pallidus and thalamus is shown in Figure 1.9. This circuit diagram portrays the dual role of the globus pallidus in setting the degree of thalamocortical excitation.

Two neurotransmitters are dominant in this circuit: the inhibitory neurotransmitter GABA and the excitatory neurotransmitter glutamate.

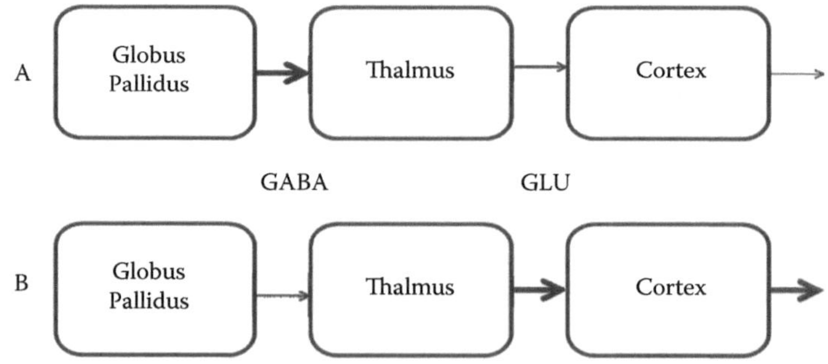

Figure 1.9 Diagram portraying the dual role of the globus pallidus in setting the degree of thalamocortical excitation. In scheme A, increased globus pallidus output inhibits thalamic activity and decreases cortical excitation. In scheme B, decreased pallidal output disinhibits thalamic excitation and increases cortical activity. Thin arrows refer to reduced activity; thick arrows refer to increased activity.

Inhibitory neurotransmitters effectively *increase* the threshold at which a neuron will fire, thus decreasing the likelihood that it will elicit subsequent neurotransmission to the target endplate (e.g., a muscle); excitatory neurotransmitters *decrease* the firing threshold and increase the likelihood of an electrochemical cascade to an endplate. Altering GABAergic neurotransmission from the globus pallidus to the thalamus by increasing or decreasing the availability of GABA at thalamic receptor sites will have a direct effect on the amount of glutamate released by the thalamus. In this example, an increase in GABA output from the globus pallidus to the thalamus inhibits thalamic glutamate. This in turn decreases cortical excitability, which would lead to a reduction in movement. Conversely, a decrease in GABA output from the globus pallidus to the thalamus causes less inhibition on thalamic glutamate. This in turn increases cortical excitability and subsequent increase in movement.

Frontal-Subcortical Neural Circuits and Motor Function

The importance of the basal ganglia in the control of fine complex movements cannot be overstated. Essentially, the entire cerebral cortex projects to the basal ganglia, which in turn funnel projections back to the SMA, the frontal cortex, and motor areas of the brain stem (Houk and Wise 1995). There is convergence from cortex to striatum and divergence back to zones in the frontal lobe. As with high-tension power or telecommunication lines, the ascending pathways from basal ganglia to frontal cortex travel as parallel circuits (Alexander, DeLong, and Strick 1986).

Several authorities on the subject have proposed hypotheses for how these parallel circuits might function together for the planning and execution of complex motor behavior (Alexander et al. 1986; Albin, Young, and Penney 1989; DeLong 1990; Cummings 1993; Houk and Wise 1995; Mink 2003; DeLong and Wichmann 2007; Turner and Desmurget 2010). Alexander et al. (1986), Albin et al. (1989), and Delong (1990) were among the earliest groups to conceptualize how the frontal-subcortical circuits might process information, particularly for motor behavior. Their functional model comprised multiple parallel circuits organized anatomically and physiologically to subserve specific motor, cognitive, and emotional behaviors. It is through this mechanism of functionally segregated circuits that willed movements are initiated from diverse cortical regions.

In its simplest form, the circuit model is based on the inhibitory functions of the basal ganglia output to the thalamus. By increasing inhibitory outflow to the thalamocortical projection sites, the basal ganglia exert a "breaking" action inhibiting the cortical motor pattern generators. Conversely, by decreasing inhibitory outflow, the basal ganglia facilitate cortical motor pattern generation. In sum, within the motor circuit, the basal ganglia function to selectively facilitate desired movement patterns and inhibit competing or undesirable patterns.

DeLong and Wichmann (2007) proposed that tonic pallidal output from the basal ganglia motor circuit to thalamocortical neurons govern the overall amount of movement. Because pallidal output is inhibitory to thalamocortical neurons, increasing pallidal output decreases, whereas decreasing pallidal output increases the overall amount of movement. The balance between increasing and decreasing pallidal outflow via the direct and indirect striatopallidal pathways ensures proper scaling and focus of movements. This balance is regulated by dopamine, which differentially facilitates or inhibits pallidal outflow by targeting striatal dopamine receptors within the direct pathway (D1 receptors) or indirect pathway (D2 receptors), respectively.

Houk and Wise (1995) derived a model of basal ganglia function based on information processing theory. In their model, spiny neurons within the striatum receive convergent signals from the cortex. The spiny neurons function as pattern classifiers. Once patterns are learned (a process that is thought to develop following repeated modulation of these spiny neurons by dopamine), the neurons can recognize familiar patterns in the input signals from the cortex. The familiar input then signals the appropriate burst pattern of spiny neuron discharge to pallidal (and subsequently thalamic) neurons to initiate or suppress ongoing cortical activity. In this model, the striatal spiny neurons possess a form of "working memory" of cortical firing patterns.

The frontal-subcortical motor circuit has been the most widely studied basal ganglia circuit, largely because of its importance in movement disorders such as Parkinson's disease and Huntington's disease and because of its relevance to a variety of neuropsychiatric disorders. Turner and Desmurget (2010) summarized decades of research on the motor and integrative functions of the basal ganglia, particularly the pallidothalamic output projections. Regarding basal ganglia regulatory functions, evidence supports a role for the refinement of ongoing motor commands rather than initiation of movement sequences. For example, the rate of firing of pallidal output neurons increases in proportion to changes in movement. As the pallidothalamic projects are inhibitory, this firing pattern would suggest ongoing inhibition of undesired movements necessary to bring about a desired movement change. Furthermore, by funneling inputs from diverse cortical regions, the basal ganglia likely integrate cognitive and motivational information with the movement kinematic plan to bring about a kinematically appropriate context-specific movement (Turner and Anderson 1997).

Prior studies demonstrated that while several important aspects of motor control are preserved following interruption to the pallidothalamic pathway (e.g., reaction time, error correction, learned motor sequences), other aspects may be severely compromised. For example, damage to GPi reduces movement velocity (e.g., bradykinesia) and causes undershooting of movement extent (e.g., hypometria, micrographia). The question remains: How is it that the failure of a "braking" system would reduce speed and movement extent? One explanation is that reduced movement velocity and movement extent are the result of antagonistic muscle co-contractions, the simultaneous contraction of opposing muscles. Disinhibiting the command to relax the antagonist muscle (e.g., the flexor carpi radialis) while allowing the agonist muscle (e.g., the extensor carpi radialis longior) to contract would reduce movement speed and extent (Anderson and Horak 1985).

One interesting function of the basal ganglia motor circuit is in regulating the speed and size of the movement—that is, the movement gain. Individuals with diseases of the basal ganglia consistently demonstrate an inability to scale the initial agonist muscle burst to meet the demands of the task (Hallet and Koshbin 1980; Caligiuri, Lohr, and Ruck 1998; Pfann et al. 2001). Compelling evidence from neuroimaging studies demonstrates a strong association between basal ganglia activation and gain adjustments during movement (Turner et al. 2003; Pope et al. 2005; Spraker et al. 2007; Thobois et al. 2007). As an independent control factor, movement gain is the optimal balance between the cost of movement (time, energy, control complexity) and the reward (see Chapter 3). Optimal attainment of a reward through a specified movement requires adjustment of costs such as velocity, movement extent, and error tolerance. This "movement gain" hypothesis is consistent with a larger view of the basal ganglia in the regulation of action motivation (Salamone et al. 2009; Turner and Desmurget 2010) when one considers that the basal ganglia funnel convergent information from emotional and sensory association as well as motor areas of the brain.

The Cerebellum and Brain Stem

In this section, we briefly review the role of the cerebellum and descending motor pathways through the brain stem in generating and maintaining precise hand motor control. These lower centers function in the reflexive and coordinative control of movement. The cerebellum functions in motor learning and the precise control of timing and accuracy by integrating sensory and motor information. The brain stem has a lesser role in motor control, acting as a relay station for all descending and ascending cranial and spinal nerves.

Cerebellum

The cerebellum is a large mass occupying a region of the brain below the occipital lobe and posterior to the brain stem. The cerebellum accounts for approximately 11% of the entire mass of the brain. It consists of a central region called the vermis and two winged lobes called the cerebellar hemispheres. The cerebellum contributes to the coordination, accuracy, and precise timing of movement. It accomplishes its task by integrating sensory input (in the form of proprioceptive, visual, and tactile sensation) with input from other parts of the brain, including the thalamus, basal ganglia, and cortex.

The cerebellum is a complex center. Internally, it contains hundreds of millions of mossy fiber cells that project onto billions of granule cells. The granule cells in turn converge onto less than 20 million Purkinje cells. Thus, as the main output structures of the cerebellum, the Purkinje cells project highly focused sensory and motor information to cortical, subcortical, and brain stem areas (Apps and Garwicz 2005). It is well positioned to play a key role in motor learning. In general, the cerebellum coordinates the kinematic parameters of movements by comparing actual movement to the intended movement and forwarding any error to the cortex for refinement. A strong network of fibers from the inferior olivary bodies (located in the brain stem near the medulla; see later discussion) to the cerebellum mediates sensory input from muscle and other peripheral receptors.

In his now classic paper, Kornhuber (1971) attributed control of preprogrammed ballistic movements to the cerebellum, while movements requiring ongoing feedback and monitoring were the provenance of the basal ganglia. Kornhuber studied saccadic eye movements (tiny horizontal ramped eye movements) in patients with cerebellar lesions. Unlike hand, arm, or leg movements, saccadic eye movements are not capable of smooth movement. When moving to a target in the visual field, the duration (and thus distance) of the saccade is preprogrammed and always ballistic. There is no error correction. In patients with cerebellar lesions, the saccadic eye movements become dysmetric; that is, distance cannot be controlled. On the basis of these observations, Kornhuber reasoned that the function of the cerebellum was to calculate the duration of the agonist muscle burst for rapid preprogrammed (open loop) movements. Based on his clinical observations and work with lesioned animals, Kornhuber hypothesized that the cerebellum functions in translating the spatial parameters of movement into time (duration of movement) for ballistic preprogrammed movements.

If the cerebellum functions in establishing the timing and duration of *ballistic* movements, why does it need sensory input from visual areas of the cortex, muscle spindles (feedback on muscle length), or joint receptors (feedback on muscle force)? One theory is that the cerebellum relies on peripheral "feedback" to inform the motor program about *starting position*. Starting positions differ for each ballistic movement, whether it is for writing a signature or visually tracking an object. In order to calculate the time needed to program the ballistic movement, knowledge of the starting position is needed. For example, when writing a stylized signature with the wrist slightly flexed, the starting position of the pen is in a different location than if the signature were written with the hand slightly extended. Sensory information from the muscles and joints of the hand reaches the cerebellum and is used in the calculation of the timing of pen movements necessary for a normal signature.

While more recent work on the function of the cerebellum has led to the expansion of its primary role in motor control to include language (Beaton and Marien 2010), cognitive behavior (Gillig and Sanders 2010), and learning and memory (Thach 1998), studies continue to demonstrate that the cerebellum system provides a clock-like timing signal to the cerebellum for the control of the temporal parameters of movement and to enable rapid error correction (Llinas 2009).

Brain Stem

We learned in earlier sections of this chapter that the basal ganglia project the majority of their output fibers back to the cerebral cortex. These highly refined neurochemical signals now must reach the muscles in the periphery to effect movement. The pathway from motor cortex to spinal column passes through the brain stem nuclei in what is referred to as the final common pathway. Literally, all descending corticospinal projections terminate within brain stem nuclei. Subsequently, the final descending motor projections en route to the spinal cord originate in the brain stem. Furthermore, all ascending sensory projections from the muscles and other sensory organs terminate on relay centers within the brain stem prior to reaching the cerebellum, thalamus, or cortical areas.

Pathways within the brain stem are topographically organized, such that the spatial representation of lower extremity to upper extremity from medial to lateral surface of the motor cortex is faithfully represented through the motor areas of the brain stem and spinal cord. The brain stem connects the diencephalon (the region above the midbrain that includes the thalamus, among other structures) to the spinal cord. The brain stem consists of white matter fiber tracts surrounding a core of gray matter and the brain stem comprises the midbrain (mesencephalon), pons, and medulla oblongata.

The midbrain is the uppermost part of the brain stem. The midbrain projects fibers to higher and lower brain centers and is responsible for maintaining visual and auditory reflexes. A key midbrain structure, the substantia nigra, provides the main dopaminergic input to the basal ganglia. Anatomically distinct nuclei within the midbrain include the origins of the third and fourth cranial nerves (for control of eye movements) and three other nuclei with important motor functions: the red nucleus (sends fibers to lower motor neurons), the substantia nigra pars compacta (projects dopamine to the striatum), and the reticular formation (part of a primary pain desensitization pathway; also involved in arousal, consciousness, and autonomic reflexes).

The pons is the bulging region in the middle of the brain stem. It is mainly a relay region, making connections with ascending and descending cranial and spinal neurons. The pons also contributes to the regulation of respiration.

The medulla oblongata is the lower portion of the brain stem that merges with the spinal cord at the foramen magnum. The medulla is the autonomic reflex center and maintains normal cardiovascular, respiratory, and vegetative homeostasis. Together with the pons, the medulla maintains consciousness and regulating the sleep cycle. Two landmark features (each with bilateral representation) characterize the medulla. These are the pyramids, where descending fibers decussate prior to innervating spinal neurons on the contralateral side of the body, and the inferior olivary bodies, which provide sensory input in the form of muscle proprioception to the cerebellum. This decussation marks the boundary between the brain stem (medulla) and spinal cord.

Summary

The goal of this chapter was to provide a fundamental understanding of the relevant neuroanatomical and neurochemical bases of motor control. Human movement is governed by neuronal activity originating and terminating at multiple levels within the central nervous system. While the cerebral cortex plays an important role in generating the initial muscle forces necessary to move a limb and integrating sensory feedback for the ongoing monitoring of muscle force, deeper brain structures such as the basal ganglia, substantia nigra, thalamus, and cerebellum ensure that movements are executed with precision and synergy.

The basal ganglia and neighboring subcortical nuclei play a key role in the maintenance and stabilization of voluntary movements, regulation of muscle tone, and integration of afferent information from the periphery. Through a complex network of circuits originating in the cortex, the basal ganglia funnel sensorimotor information back to the cortex to complete a massive feedback loop. The interaction of excitatory and inhibitory neurotransmitters throughout the basal ganglia balances and tunes the network to ensure optimal motion control. Building on the basic understanding of the neuroanatomical and neurochemical bases of motor control, we turn our attention in subsequent chapters to the cortical and subcortical control of handwriting movements.

Neuroanatomical Bases of Handwriting Movements

2

Introduction

Having reviewed the fundamental neuroanatomy and neurochemistry of hand motor control in Chapter 1, we can now turn our attention to what we know about brain anatomy and chemistry for the control of handwriting. Information is available from two general sources: lesion studies of neurological patients and functional neuroimaging involving generally healthy writers. Lesion studies are typically case reports or case series of patients who experienced cerebral vascular accidents (stroke) or developed brain tumors and underwent surgical excision.

Lesion Studies

In this section of the chapter, we attempt to integrate research on handwriting following brain lesions into a generalized understanding of the relationship between specific brain regions and the execution and ongoing monitoring of handwriting. Several reports linking specific cortical areas to handwriting have appeared in the published literature. Most of these are case reports of patients recovering from a stroke (vascular accident) or surgical removal of brain tumors. Collectively, they reveal broad representation throughout the frontal and parietal lobes and the basal ganglia for handwriting.

Handwriting Change Following Vascular Accidents

Cerebral vascular accidents or strokes are the most common form of brain injury in adults. Strokes result when the blood supply to the brain is interrupted either by blockage (thrombotic or embolic strokes) or bursting of a blood vessel (hemorrhagic stroke). Strokes can occur in any region of the brain. Brain stem strokes are generally fatal; while strokes to subcortical or cortical areas are survivable, they can leave the individual with impaired function. The study of individuals with residual impairments to cognitive or motor function following stroke offers serendipitous opportunities to advance our understanding of brain–behavior relationships in a natural setting including the study of handwriting movements.

Subcortical Vascular Accidents

Subcortical strokes are not common and, when they occur, functional impairments often remit within weeks. We found three published reports involving individuals experiencing strokes restricted to subcortical regions including the thalamus (Kim et al. 1998; Ohno et al. 2000) and striatum (Nakamura et al. 2003) leading to handwriting impairment. Kim et al. (1998) described a right-handed patient who presented with micrographia (abnormally small letters) as his only motor sign following a subcortical stroke. Strength and sensation were normal as were fine motor skills involving the upper extremities. When asked to write, he held the pen normally and initiated writing movements with normal speed. The handwritten samples were smaller in amplitude than prestroke samples and, as he continued to write, the letters became smaller and more disorganized. Progressive decrease in the size of handwritten words or numbers during continuous handwriting is characteristic of Parkinsonian micrographia. Closer inspection of the anatomical brain scans from this patient revealed abnormalities in the left thalamus.

Functional brain imaging performed to evaluate the distribution and binding of dopamine to various regions throughout the basal ganglia showed decreased dopamine in the left striatum. Interestingly, this patient's stroke did not involve the striatum directly. However, since dopamine from the substantia nigra projects to the striatum, it appeared in this case that the initial thalamic lesion interrupted communication with centers downstream that modulate dopamine release and caused a decrease in nigrostriatal dopamine transmission resembling parkinsonism. This case underscores the complex interactions among various nuclei within the basal ganglia and how damage to one area (e.g., the thalamus) can affect function of another (e.g., the substantia nigra) and lead to altered handwriting.

Whereas Kim's thalamic patient experienced micrographic handwriting, Ohno et al. (2000) described a patient with a thalamic stroke who presented pure apraxic agraphia. Apraxic agraphia refers to an inability to sequence letters when writing, but not other forms of verbal expression, that is typically seen following left frontal cortical damage. This patient's handwriting was characterized by omissions and additions of letters; micrographia was not reported. Together, the Kim et al. and Ohno et al. case reports support a complex role of the thalamus as an intermediate nucleus for the execution of handwriting behavior.

When neural projections between the thalamus and striatum are disrupted by a thalamic stroke, handwriting could become micrographic; this is typically observed with loss of striatal dopamine such as in Parkinson's disease (PD; to be discussed in greater detail in Chapter 10). However, when neural projections between the thalamus and cortex (e.g., supplementary motor area) are disrupted by a thalamic stroke, handwriting takes on more of an apraxic disorganized form. Thus, the thalamus appears to function as an intermediate relay station with projections to both lower level centers, such as the basal ganglia, and higher cortical centers to mediate multiple aspects of handwriting behavior.

Nakamura and colleagues (2003) reported a patient who suffered a stroke affecting his left basal ganglia—specifically, the putamen. As predicted by the Kim et al. case, the Nakamura et al. case also exhibited micrographia. Unlike the Kim et al. case, this patient exhibited mild reflex signs indicative of frontal lobe damage, yet brain imaging studies revealed normal appearing frontal lobes. Two important aspects of brain function and handwriting are revealed by this case. First, the putamen, a major nuclear region of the striatum, appeared to have an important role in the handwriting impairment. As noted before, micrographia is a hallmark sign in Parkinson's disease. Whereas the pathology in Parkinson's disease originates in the substantia nigra and leads to a loss of dopamine neurotransmission to the putamen, the Nakamura et al. case implicates the putamen in the nigrostriatal pathway underlying micrographia. Second, the co-occurrence of micrographia with frontal release signs suggests a functional pathway linking the striatum and frontal lobe.

These published cases on handwriting characteristics following vascular lesions to subcortical brain regions shed light on the importance of the basal ganglia in handwriting. They contrast with the traditional viewpoint that the programming and control centers for handwriting are the provenance of higher cortical areas of the brain, reflecting the voluntary and linguistic roles of this uniquely human function.

Findings from individual case reports and case series on the effects of stroke and surgical procedures for tumor resection (see below) demonstrate that cortical lesions can produce two forms of agraphia: spatial agraphia (Ardila and Roselli 1993) and apractic agraphia (Alexander, Fischer, and Friedman 1992). The term "apractic agraphia" refers to a specific condition characterized by deteriorated handwriting in the presence of normal sensorimotor function, cognitive, and language abilities. Apractic agraphia has been described as "loss of motor programs to form graphemes" (Roeltgen and Heilman 1983), impairment to the "graphemic area that generates physical description of letter" (Crary and Heilman 1988), or "a selective impairment of the execution of writing sequence, manifested as an abnormal order of writing strokes" (Otsuki et al. 1999).

Spatial agraphia is usually associated with lesions to the right hemisphere, whereas in apraxic agraphia, the left hemisphere is involved. Alexander et al. (1992) describe spatial agraphia as having "margins [that] are unformed or incorrect…adjacent letters are in incorrect apposition and letter strokes are reiterated…the written line wanders off the horizontal" (p. 248). Handwriting in apraxic forms of aphasia is generally laborious and clumsy; letters are often written out of sequence, yet decipherable. Alexander and colleagues noted that apractic agraphia likely represents various degrees of dysfunction of the overlearned handwriting motor program.

Cortical Vascular Accidents

In this section, we review findings from published studies on the consequences of cortical vascular lesions on handwriting. One of the challenges in drawing conclusions about the relationship between normal brain functioning and handwriting from research on lesions to the left frontal or parietal lobes (key areas thought to be important in storing and executing the graphomotor program) is that these regions also govern written and oral linguistic processes. Thus, dissociating the motor from the linguistic aspects of handwriting is confounded by the dual roles these cortical areas have in expressive language.

A good example of this problem is the study by Basso, Taborelli, and Vignolo (1978). These researchers reviewed the records of 500 adult patients with left brain damage due mostly to vascular lesions. They were interested in identifying whether pure handwriting deficits (i.e., in the absence of language impairment) could be dissociated from the more common language disorders that accompany most left hemisphere strokes. They found only two cases of pure agraphia. While extremely rare, the lesions in these cases were both located in the left superior parietal region. The authors described the handwriting of these two patients as "awkward and trembling…with occasional additions of a few loops and curves to letters…minimal spatial distortions" (p. 559). Nonetheless, this study is useful as it reveals specific cortical regions underlying the nonlinguistic motor aspects of handwriting.

Evidence in support of specific cortical involvement in handwriting is based on case reports of patients surviving strokes to the cerebral cortex (Valenstein and Heilman 1979; Auerbach and Alexander 1981; Roeltgen and Heilman 1983; Crary and Heilman 1988; Levine, Mani, and Calvario 1988; Alexander et al. 1992; Otsuki et al. 1999). Auerbach and Alexander (1981) reported an interesting case with impairment of visually guided hand movements with pure (motor) agraphia. The individual suffered a clot in a vessel supplying blood to the left superior parietal lobe, destroying a small portion of the region of his brain associated with Brodmann's area 7. The patient's chief complaint following the stroke was difficulty with handwriting, especially when attempting to sign his name. Auerbach and Alexander described his handwriting as "untidy and poorly formed…the patient would cross lines or make loops in inappropriate places...present in cursive and printed productions" (p. 431).

Alexander et al. (1992) describe a patient with a lesion to his left superior parietal lobe following a stroke who exhibited impaired handwriting that persisted several weeks following the stroke. The patient's handwriting was laborious, often requiring several minutes to complete a single word. Unlike the Auerbach case, Alexander's case also exhibited difficulty performing visuospatial tasks with the right hand, although her language skills were relatively intact. Otsuki et al. (1999) reported a case of pure agraphia following a hemorrhagic stroke in the left superior parietal lobe that completely recovered within 1 month following the vascular episode. Upon examination the patient was quoted as saying, "My hand slips, although I know how to write" (p. 234).

When cortical lesions interrupt handwriting, they tend to be limited to two small regions: the left posterior frontal lobe and the left superior parietal lobe. Pure agraphia following cortical vascular lesions is rare as most patients who have handwriting impairment also have difficulty with language expression (aphasia), complex motor tasks involving the hand (apraxia), or visuomotor control. The degree to which these other problems are observed is directly related to the size and depth of the brain lesion. Smaller lesions to the left posterior frontal lobe or left superior parietal are more likely to manifest as pure agraphia than larger lesions.

In summary, the clinicoanatomical literature on localizing the brain's control center for handwriting is limited. Evidence supporting a key role of focal cortical areas comes largely from fewer than a dozen case reports. In an integrated review of the literature prior to 1990, Alexander et al. (1992) concluded that

> Despite markedly different assessment methods, all the reports coalesce around a single theme. There is a region in the parietal lobe, usually but not invariably the language-dominant one, that directs the capacity to generate the learned motor patterns of writing in a facile, automatic manner. The region is apparently dorsal, in or around the junction of the superior angular gyrus and the superior parietal lobule. After lesions in this region, the sequence of movements for writing cannot be activated despite knowledge of letters, knowledge of how words are spelled, and normal sensorimotor function. (p. 250)

Handwriting Change Following Surgical Resection for Brain Tumor

Unlike the opportunistic observations from patients who suffer from vascular injury to critical brain regions, the surgical removal of tissue from these and neighboring brain areas during surgical resection to treat brain tumors offers a more systematic and prospective approach to understanding the neuroanatomy of handwriting. The neurosurgical literature contains three recent case series describing specific handwriting changes following removal of cortical tissue from frontal and parietal areas (Lubrano, Roux, and Démonet 2004; Scarone et al. 2009; Magrassi et al. 2010). The goal of these surgical interventions was to identify and spare surrounding language areas to the extent possible during tumor excision. Awake patients are asked to perform various cognitive and motor tasks, including handwriting, while surgeons probe nearby cortical tissue to map functional boundaries. It is from this approach that handwriting-specific sites have been identified in the frontal and parietal lobes.

Lubrano et al. (2004) published results from 14 surgical patients with tumors located in the left or right frontal gyri or rolandic fissure. Figure 2.1 shows the cortical maps of the probes eliciting changes in handwriting. They found that handwriting was interrupted during direct stimulation in the dominant inferior and middle frontal gyri. In many cases, the handwriting interruption occurred in the absence of other expressive language interference, suggesting a pure motor rather than language-based impairment. Electrical probing to the superior frontal gyrus yielded no specific writing errors, whereas stimulation to the middle and inferior frontal gyri resulted in

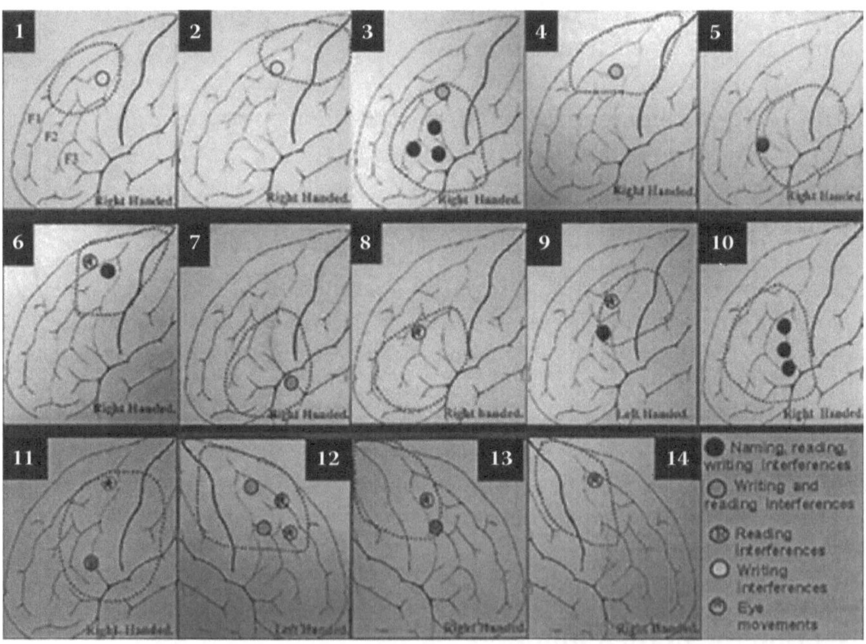

Figure 2.1 Cortical maps from 14 tumor patients showing regions where electrical probes elicited changes in handwriting. (From Lubrano, V. et al. 2004. *Journal of Neurosurgery* 101:787–798. With permission.)

writing arrests or perseverations (predominantly middle frontal gyrus), letter substitutions (middle and inferior fontal gyrus), and illegible script (predominantly inferior frontal gyrus).

The authors noted that the attribution of a writing-specific interruption to a specific frontal gyrus was highly variable among patients and that many patients exhibited various combinations of writing, reading, and speaking interruptions following stimulation to the same gyrus. However, the authors concluded that a specific motor handwriting deficit could occur with a small lesion that damages the writing area in the middle frontal gyrus but spares nearby language areas.

In a similar study, Scarone et al. (2009) analyzed postoperative results from 15 patients who underwent surgical excision of the supplementary motor area (SMA), middle and inferior frontal gyri (Brodmann area 6), or the superior parietal lobe to treat glioma. A summary of their findings is presented in Table 2.1.

Handwriting examples and lesion characteristics from one of the Scarone et al. (2009) series of cases are shown in Figure 2.2. Postoperatively (B in Figure 2.2), the handwriting for this patient was very irregular and trembling. Many letters and words were difficult to recognize.

The Scarone et al. (2009) study underscores the complexity of the anatomofunctional network for handwriting. Several cortical sites appear to form a network that subserves the spatial, motor, and linguistic aspects of handwriting. This network has at least five zones including the superior parietal lobe, the supramarginal gyrus, the SMA, a zone capturing the middle and inferior frontal gyri, and the insula. The linguistic and motor functions within this network can be dissociated. Specifically, the language subcomponent of this network likely resides in the middle and inferior

Table 2.1 Summary of Findings on Handwriting Impairment Following Surgical Excision. BA indicates Brodmann area.

Lesion Site	Handwriting Characteristics
Left superior parietal lobe (BA 7)	Spatial agraphia: spatial disorganization; hesitant, shifted to the right side of the page; disturbed spatial array of letter sequences; difficult writing on a horizontal line; perseveration
Left (or right) supramarginal gyrus (BA 40)	Apractic agraphia; slow and perseverative
Left (or right) SMA	Effortful, irregular, and trembling; letters often unrecognizable
Left middle and inferior frontal gyrus	Minor impairment of handwriting; some hesitation; normal size or shape of letters
Left insula	Writing errors of substitution and repetition

Source: Scarone, P. et al. 2009. *Surgical Neurology* 72:223–241.

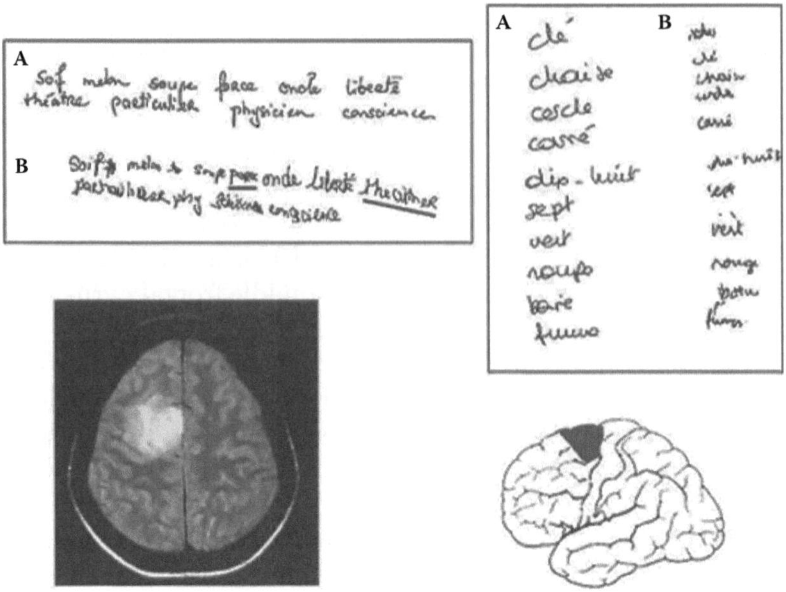

Figure 2.2 Examples of handwriting from a single patient for sentence production (top left) and words to dictation (top right) before (A) and after (B) surgical removal of a tumor encompassing the SMA and areas 4 and 6 of the left hemisphere. Lower plates show lesion location. (From Scarone, P. et al. 2009. *Surgical Neurology* 72:223–241. With permission.)

frontal zone, while the motor subcomponent includes the superior parietal lobe and the SMA.

A recent case report by Magrassi et al. (2010) lends further support for a complex handwriting network with both linguistic and motor components. These investigators were able to selectively induce and reverse language and motor interruptions for handwriting by stimulating frontal and parietal areas close to the margin of a tumor. Electrical stimulation within the superior parietal lobe interrupted the handwriting while stimulation near frontal lobes induced lexical errors. Based on their findings, the authors proposed that cortical control of handwriting involved a central linguistic process that converges onto the peripheral motor process within the superior parietal lobe. Such convergence is entirely consistent with the notion of a handwriting network proposed by Scarone and colleagues.

Functional Neuroimaging Studies

Since antiquity, the understanding of the relationship between brain anatomy and behavior has come from careful observation and assessment of individuals who sustained damage to specific areas of the brain. While a great deal has been learned about handwriting from this research, several problems raise questions as to whether the clinicopathological approach is a valid model for studying such complex motor behavior as handwriting. For example, a brain lesion anywhere within the "handwriting network" could alter the normal behavior of this network. This is akin to a familiar scenario in which a traffic jam on a busy roadway will lead to congestion on an alternate or parallel road as drivers seek alternate routes to avoid the jam.

We have summarized evidence from previous studies of individuals with damage to the thalamus that exhibit apractic agraphia, a condition thought to stem from lesions to higher cortical areas such as the superior parietal lobe or supplementary motor area. Another problem is that brain trauma, whether from a stroke or surgical intervention, produces edema. This swelling could transiently alter functioning of otherwise healthy brain tissue, with recovery varying widely among individuals.

Within the past decade, several groups have begun to utilize functional neuroimaging techniques such as magnetic resonance imaging (MRI) or positron emission tomography (PET) to map the neuroanatomy of motor behavior among normal healthy individuals. Of particular relevance here are several studies that shed light on the location and dynamics of a "handwriting center" in the normal brain (Menon and Desmond 2001; Siebner et al. 2001; Sugihara, Kaminaga, and Sugishita 2006).

Functional MRI (fMRI) relies upon the change in the electromagnetic properties of oxygenated blood as it flows through vessels to areas purported to be active during behavioral (e.g., motor) tasks. As active brain areas demand greater oxygenated blood flow than inactive areas, during the few seconds needed to attain this increase in blood flow the magnetic polarity of the molecules in hemoglobin flips. This polar flipping is detected by the MRI scanner and can be quantified by software. This is known as the BOLD (blood oxygen level dependent) response and has been perfected to generate high-resolution images of near real-time "activation"; these are then aligned onto a structural anatomical map of the brain.

In a typical experimental paradigm, stimuli are presented (visually or auditorily) to the subject in a systematic "on–off" manner while the scanner is continuously collecting data. The subject is instructed to respond (e.g., writing on a tablet, pressing a keyboard, etc) when the stimulus is "on", and then to rest or perform a neutral task during the "off" condition. The BOLD response is then analyzed statistically for patterns that coincide with the on–off pattern of the stimuli. A tightly coupled BOLD response in one area of the brain is used to infer that that brain region is "active" for that particular behavior.

Menon and Desmond (2001) employed fMRI to help localize critical brain areas active during writing to dictation in 14 healthy right-handed subjects. The task consisted of 12 40-second trials during which subjects wrote sentences to dictation alternating with 40 seconds of passive fixation. The investigators then identified clusters of brain space having BOLD responses that coincided with the writing task. They found four areas associated with writing: the superior parietal lobe (SPL; Brodmann area 7), the inferior parietal lobe (Brodmann area 40), the supplementary motor area (Brodmann area 6), and the sensorimotor cortex (Brodmann areas 1, 2, 3, and 4) of the left hemisphere. The finding that the SPL was particularly active during the writing task is consistent with the lesion studies showing impaired or interrupted writing with damage or electrical stimulation to SPL. Interestingly, the SMA and motor cortex of the left hemisphere are known to play key roles in motor programming (feedback correction) and execution and were found to be active during handwriting in this study as well.

In a slightly more complex experiment involving fMRI, Sugihara et al. (2006) attempted to dissociate cortical areas associated with writing from those associated with naming. The investigators hypothesized that common areas in the left and right hemisphere would be critical for handwriting and not active during silent naming of the same words. Their results from 20 right-handed healthy subjects showed three common areas to be consistently active during writing regardless of whether subjects wrote with the left or right hand. These included the anterior limb of the left supramarginal gyrus (Brodmann area 40), the left SPL (Brodmann area 7), and the left superior frontal gyrus (Brodmann area 6, or the premotor area). The Sugihara et al. findings are consistent with those of Menon and Desmond (2001) and underscore the importance of Brodmann areas 40 and 7 in the central processing of handwriting movements. They show that the left frontal region (the premotor area) may be a key player in this process.

PET functions differently. In this technique, a scanner detects a radioactive tracer that is injected into the individual. Once transported to the brain, the tracer binds to certain molecules, usually glucose (but other tracers bind to different molecules or receptors, such as dopamine). The researcher can then obtain a visual map of glucose metabolism (or change) anywhere in the brain. This functional technique quantifies brain activity not in terms of blood flow (as with MRI), but rather in terms of glucose metabolism or uptake by the receptor of a radioactive tracer (specific to a neurotransmitter). These "activity" maps are then aligned with higher resolution individualized anatomical maps to localize areas of brain activity associated with a behavioral task. PET is used to localize brain activation based on increase in glucose or increase in the molecules (such as dopamine) binding to nerve receptors.

Sieibner et al. (2001) used PET to examine differences in brain activity during open loop handwriting (fast, without feedback or monitoring) compared to closed loop handwriting (slow, requiring ongoing self-monitoring). To verify whether subjects were performing the open loop (fast ballistic) or closed loop (more slowly modulated) task, kinematic analyses were performed on the digitized handwriting samples. Pen movements with only a single velocity peak per stroke were considered open loop; pen movements with multiple velocity peaks indicated changes in speed based on internal feedback and were considered closed loop. Thus, the number of velocity peaks per stroke served as the independent variable representing degree of internal monitoring of handwriting speed.

PET results revealed strong correlations between cerebral blood flow in the left SMA and right precuneus (the mesial extent of Brodmann's area 7 or the SPL) and number of velocity peaks per stroke. These findings indicate that the left SMA and right SPL are particularly involved in generating closed loop writing movements. The Siebner et al. findings provide empirical support for the SMA in the sensorimotor integration during execution of fine hand movements, including handwriting. Interestingly, as with prior studies (Seitz et al. 1997; Ibanez et al. 1999), Siebner et al. failed to find PET activity within any of the basal ganglia regions that corresponded to any kinematic variable during the open loop handwriting. This may be due to technical or resolution limitations. On the other hand, closed loop velocity-controlled handwriting movements were associated with activation of the basal ganglia.

Summary

Convergent findings from lesion, neurosurgical, and functional neuroimaging research support the existence of a network of cortical and subcortical regions that govern handwriting movements. This network has at least five zones, including the SPL, the supramarginal gyrus, the SMA, a zone capturing the middle and inferior frontal gyri, and the insula. Linguistic and motor functions within this network can be dissociated. Specifically, the language subcomponent of this network likely resides in the middle and inferior frontal zone, while the motor subcomponent includes the SPL and the SMA. It is not surprising that the left hemisphere would house this important function because of the role it plays in language and the overlap between linguistic and graphomotor behavior in humans.

Case reports of patients surviving vascular accidents involving the basal ganglia confirm the importance of the striatum (especially the putamen) in the ongoing monitoring of handwriting movements. Such individuals exhibit impairments in handwriting that resemble PD. However, unlike PD, micrographic handwriting following a basal ganglia stroke is transient, usually disappearing within weeks following the stroke.

Cortical-subcortical circuits integrate sensory information from the periphery with motor commands for on-line monitoring of hand movement. An important structure in this feedback circuit is the SMA. The SMA receives projections from the basal ganglia and posterior sensory cortices to integrate sensorimotor information during ongoing execution handwriting. As such, the SMA serves as a comparator in a feedback circuit that begins in the SPL, passes through the basal ganglia, and, along with sensory information from the periphery, merges onto the SMA where information is uploaded before being passed back to the SPL. If the SMA is involved in motor tasks requiring internal monitoring, one could hypothesize that activation of the SMA (observed using fMRI or PET imaging) would differ when a person is producing a forged (simulated) signature versus his or her own genuine signature. Such an experiment would validate the importance of the SMA in the ongoing monitoring of handwriting.

Models of Handwriting Motor Control

3

Introduction

The aim of this chapter is to provide an overview of the various models of motor control, their empirical support, and application to handwriting motor control. The processes involved in executing natural handwriting movements are extraordinarily complex and involve multiple brain regions working in concert to bring about coordinated, precisely timed, multijoint movements capable of adapting to ongoing and often unpredictable environmental constraints. As a first step in understanding this process, scientists must reduce its complexity into simpler manageable units or models. These manageable units can be visual, mathematical, or computational and can be used to develop testable hypotheses on the nature of motor behavior under various conditions.

It is important to distinguish a theory from a model. A theory is a plausible general principle or body of principles offered to explain a phenomenon or a prediction based on previous observations or experiments. The 2005 edition of *The American Heritage Dictionary* defines a model in scientific applications as "a systematic description of an object or phenomenon that shares important characteristics with the object or phenomenon." That is, a model is a simplified system that illustrates or exhibits the same behavior as the more complex system. Theories are not testable whereas models *enable* testable hypotheses. While no single model can be expected to represent all aspects of a complex motor behavior such as handwriting, some models are better capable of generating testable hypotheses than others.

Many of the models discussed in this chapter were introduced into the motor control literature in the late 1970s and early 1980s. For an excellent historical summary of motor behavior models, the reader is referred to Abernethy and Sparrow (1992). Prior to the 1990s, the more robust models of motor behavior were derived from fundamental theories such as the closed loop or feedback theory of motor control (Adams 1971), motor programming or schema theory (Keele 1968; Schmidt 1975), impulse-variability theory (Schmidt et al. 1979), and dynamical or oscillatory theories (Kelsoe et al. 1981; Kugler and Turvey 1987). More recent computational models were motivated by inconsistencies between the theoretically based models and real-world observations. Provocative computational models are based on the notion that motor behavior is programmed and executed to achieve the greatest amount of flexibility and accuracy with minimal cost in energy (cost minimization models).

These models were developed not necessarily for the purpose of understanding handwriting, but rather for advancing our appreciation of how the human nervous system controls complex movement and for the purpose of developing automated robotic systems capable of performing human-like movement (e.g., Hollerbach 1981). While the translation of these general models of motor control to the specific problem of handwriting is often incomplete, there is sufficient overlap between handwriting movements and other complex movements, such as speech, typing, and other highly programmed movements, that these models have been successful in generating testable hypotheses and applied research in the areas of rehabilitation, movement disorders, and now forensics.

When considering the strengths and limitations of a particular model of handwriting control—whether the model is derived from the physical and geometric constraints of the neuromuscular system (e.g., mass-spring model) or the assumption that efficient use of energy is a priority when using the motor system to interact with the environment (cost minimization models)—all models of handwriting control start with the assumption that a generalized program for handwriting exists. This chapter provides an overview of the few models that have enjoyed some success in advancing our understanding of normal and pathological handwriting movements. Several scholarly works on the principles of motor control in handwriting have been previously published (Thomassen and van Galen 1992; Teulings 1996; Plamondon and Djioua 2006; Rosenbaum 2010). This chapter represents only an overview of relevant models that have enhanced our understanding of handwriting. Readers with an insatiable thirst for knowledge on this subject are encouraged to explore these treatises.

Handwriting as a Motor Program

What is a motor program? Keele (1968) defines a motor program as "a set of muscle commands that are structured before a movement sequence begins, and that allow the entire sequence to be carried out uninfluenced by peripheral feedback." Keele and Summers (1976) maintain that a fundamental component of skilled motor behavior begins with the sequencing of discrete movements (also known as stimulus–response or S–R chaining). The central representation of this sequence then becomes a motor program.

A motor program is a theoretical memory structure capable of transforming an abstract code into an action sequence (Schmidt et al. 1979). For some motor behaviors, complex motor sequences are organized and controlled using a fixed set of commands timed in such a way that movement parameters such as torque, trajectory, speed, and distance may be reliably repeated. These simple motor programs develop following extensive repetition and learning and are not easily generalizable to other movements. One example of a simple motor program is the repetitive, consistent execution of swinging a golf club. Once programmed, the adept golfer can rely on the same set of action sequences to produce invariant results nearly all the time.

With regard to handwriting, Thomassen and van Galen (1992) noted that the high degree of consistency in the form of an individual's script when written using different limbs offers compelling evidence in support of an abstract motor program. The existence of a motor program presumes that the movement parameters for handwriting are not stored as discrete instructions to specific muscles, but rather as a general spatial code representing the final motor output attainable under a variety of physical or environmental constraints.

An ongoing debate in the motor control literature has been whether sensory feedback is necessary in the execution of learned motor behavior and how the central motor program utilizes such feedback. Keele and Summers (1976) argue that feedback is not necessary in a central motor program. Very rapid movement sequences, such as speech, handwritten signatures, and playing a musical instrument, can be executed without feedback (i.e., open loop). The notion that open loop motor control is a key element of a motor program was a popular concept in the late 1960s and 1970s following a series of electrophysiological studies on motor behavior in nonhuman primates. Keele (1968) and others (Glencross 1977) observed that the interval between movement sequences during rapid skilled movements was too short to make use of kinesthetic feedback, thought to require an approximately 100 ms delay.

However, more sophisticated research by Evarts and Tanji (1974) revealed that the sensory-motor kinesthetic feedback loop could be realized in less than 50 ms. These results challenged the time-delay hypothesis of open loop motor control. While the time-delay hypothesis was losing favor, a more convincing argument for open loop motor control was emerging (see reviews by Hinde 1969; Keele and Summers 1976). Researchers were studying the effects of complete sensory loss on movement sequencing by observing deafferentiated experimental monkeys. In this preparation, the animal no longer had access to peripheral feedback. Yet, learned movement sequences were executed in a relatively normal pattern. It could be argued, therefore, that learned movement sequences might be represented in a central motor program as a series of discrete movements encoded without regard to peripheral feedback. It was argued at the time that feedback appears to have a limited role in ongoing error monitoring for corrective actions (Keele and Summers 1976).

Nonetheless, the notion of a central motor program made up of a chain of discrete movement sequences that are faithfully executed independently of peripheral feedback remains problematic. The sequencing of discrete movements as a major feature of skilled motor behavior is at odds with the rich variability and adaptive capability known to exist in speech production (Abbs and Winstein 1990) and handwriting (van Galen and Weber 1998). An argument can be made that complex, rapid motor sequences require a program with flexibility and accommodation to guide the planning and execution of a broad variety of movements to accomplish the same goal.

A flexible motor program is often referred to as a generalized motor program (Schmidt and Lee 1999). Unlike simple motor programs, both variant and invariant movement parameters are coded in a generalized motor program. Generalized motor programs account for the ability to achieve the same movement outcome with different muscle groups (a concept referred to as motor equivalence, which will be discussed later in this chapter). Interestingly, the most compelling evidence for the existence of a flexible and adaptive generalized motor program comes from empirical research on handwriting.

The consideration of handwriting as programmed motor behavior can be somewhat problematic. For one, it can be readily observed that handwriting is a serial motor behavior with individual letters making up words and words making up sentences in series. Because individual strokes are rarely produced faster than 80 ms (Teulings and Thomassen 1979), Thomassen and van Galen (1992) contend that there is sufficient time for the motor program to retrieve and unpack discrete sets of instructions in real time for the *entire* series without interruption in normal fluency. This precludes the need for advanced preparation and storage of motor information in a memory buffer for serial retrieval. While there is empirical support for the subprogram retrieval model for speech production (Sternberg et al. 1978), evidence does not support a similar process for handwriting (Hulstijn and van Galen 1988). Unlike speech production, the low production rate of handwriting allows for the real-time preparation rather than subprogram buffering of movement elements.

Van Galen and Weber (1998) challenged the traditional notion that a motor program represents the discrete prestructured parameterization of muscle contraction sequences and argued for the view that the stored motor program for sequences of muscle contractions or movement actions allows for continuous and dynamic adaptation to environmental or spatial constraints. For example, Thomassen and Shomaker (1986) observed that the form of handwritten letter strokes varied with changes in the form and size of the surrounding letters and strokes. Others manipulated the size and speed of the writing task or imposed sudden unpredictable changes in the scaling of the feedback and found rapid systematic adaptations in stroke kinematic features (Wright 1993).

To further identify the constancy of the handwriting motor program, van Galen and Weber (1998) investigated the timing of letter strokes during abrupt, unpredictable shortening or lengthening of the writing space. They recorded changes in the form and kinematics of handwriting associated with rapid changes to the length of the writing line. It was hypothesized that if the motor program for handwriting involved the constant parameterization of kinematic features, the timing of letter strokes but not movement trajectory would be affected by these spatial manipulations. Conversely, if the motor program were more continuous and flexible in nature, then one would observe changes in spatial goal trajectory and not timing of letter formation just prior to the change in the length of the writing line.

Several findings from this elegant study shed light on the nature of the handwriting motor program. First, the investigators found that stroke size adaptations to writing line length manipulations occurred rapidly and continuously. Second, upstrokes were affected more than downstrokes by this manipulation, suggesting greater flexibility in the motor program than previously anticipated. Third, while upstrokes increased and downstrokes decreased in vertical stroke size, stroke duration was unchanged by this manipulation. Thus, the handwriting motor program appears to be flexible and continuous rather than prestructured and constant, and it seems to code abstract goal trajectories rather than discrete instructions for timing of muscle contractions.

There is little doubt that in the adult, handwriting is an expression of highly adaptive learned programmed movement sequences. For one, the program itself can be transferred to different muscle groups (e.g., hand to foot). Second, the program can adapt to environmental constraints. Third, there is support to demonstrate that certain features for handwriting, such as stroke duration and movement patterns, are somewhat constant despite variability in the task demands, suggesting program invariance.

The presence of a motor program for handwriting could provide many advantages to the individual. For example, an effective motor program could reduce the demands on the nervous system when executing complex movement sequences. Second, by stringing together discrete movement sequences into one programmable unit, the demands on the cognitive and memory systems are also reduced. Finally, reliance upon a stored program for executing a sequence of handwriting movements enables the writer to reallocate attention and effort to the environment for the purpose of multitasking.

As noted earlier in this chapter, models enable the development of testable hypotheses. One overarching hypothesis is that a generalized motor program for handwriting exists. However, the exact nature or elements of the handwriting program remain largely uncertain. That is, what does the handwriting program look like and how is it organized? Computational models were developed to help elucidate the nature of the handwriting motor program. In the following sections, we present overviews of three relevant and widely published models designed to simplify the complex nature of handwriting including the hierarchical model, cost minimization model, and equilibrium point model.

Hierarchical Models of Handwriting Motor Control

Hierarchical models of motor control stem from cognitive approaches to understanding motor control. Hierarchical models account for both invariant (low level) and adaptive (high level) aspects of handwriting. The hierarchical view of motor control is based on processes or subprocesses organized in a top-down manner. Hierarchical models use a top-down structure in which higher centers control or inhibit activity of the lower centers. In these models, control of motor output is decentralized with linearly distributed control assigned to multiple levels within the hierarchy. Essentially all aspects of movement planning and execution are the sole responsibility of one or more higher level (e.g., cortical) centers.

Within a hierarchical model, sequences of movements can be more easily learned and programmed because the organism does not need to learn every element of the sequence. Rather, the learned program need only consist of a single high-level motor command, which would, in turn, select a set of intermediate level motor commands; the whole movement sequence therefore would evolve in the correct order. Thus, a single high-level motor program can select a set of lower level motor elements such as trajectory, speed, etc.

Such a hierarchical network can govern completely novel motor sequences and provide one solution to Lashley's problem of serial order in behavior (Stringer and Rolls 2007). Lashley (1951) was perhaps the most famous proponent of this hierarchical cognitive approach to planning and executing movement sequences, which he referred to as "the problem of serial order in behavior." Prior to Lashley, researchers modeled the control of sequential motor behavior as a reflex chain (see Rosenbaum et al. 2007 for review). In that scheme, stimulation caused by movement x triggers movement $x + 1$, which in turn triggers movement $x + 2$, and so on in a reflexive manner. Lashley's problem with the reflex chain model was that movements can be executed faithfully even when sensory feedback is altered, rendering the "triggers" somewhat unnecessary. Furthermore, Lashley noted that some movement sequences occur in too brief a time for subsequent elements of the movement sequence to utilize feedback from the preceding elements.

Hierarchical models of motor control include a number of desirable features that enable formal testing of their validity. These models are flexible in that different movement parameters may be added or removed from the general motor program. Conversely, different movement parameters can be applied to a program for the same class of actions. Hierarchical models permit changes in the general program in response to sensory feedback or environmental change. Lastly, interactions with the environment involve the lower level elements and need not alter the higher level program itself. Repeated interaction, however, can lead to learning and updating of the higher level elements.

Hierarchical Models and Handwriting

Evidence that handwriting may be governed by a modular hierarchical organization has been the topic of research for over 40 years (Van Nes 1971; Ellis 1982; van Galen and Teulings 1983; Margolin 1984; van Galen 1991; Teulings 1996). Teulings (1996) proposed three models of handwriting: macroscopic, microscopic, and computational. According to this scheme, macroscopic models have several serial and parallel modules. Support for macroscopic models comes from studies on "slips of the pen" and effects of neurological damage. Microscopic models pertain to storage of the graphic motor pattern. They represent the motor program for the timing and trajectory of movement sequences.

According to Teulings (1996), the modular macroscopic models account for the bulk of the handwriting system. One fundamental component of this modular approach to modeling handwriting is its flexibility. That is, the initial sequence of motor commands to produce a written sentence can be transmitted to any set of muscles. As noted before, this phenomenon is often referred to as motor equivalence. Coined by Lashley (1931) and further developed by Bernstein (1947, 1967) and others (e.g., Keele 1981), motor equivalence accounts for movement sequences that can be executed by different effectors. Motor equivalence suggests that motor acts are encoded in the central nervous system as abstractions rather than specific commands or strings of commands (Wing 2000). Under motor equivalence, the motor program is unrestrained by a particular limb or muscle/joint assembly usually employed to execute a complex movement. For example, the same pen stroke can be realized by an infinite number of joint rotation patterns. Figure 3.1 portrays the classic example of motor equivalence for handwriting published by Raibert (1977).

Compelling support for the concept of motor equivalence as a motor control strategy comes from an interesting study by Rijntjes et al. (1999). These investigators utilized functional magnetic resonance imaging (fMRI) to test whether performing a writing task with different effectors activated the same cortical regions. Activation of overlapping cortical areas during writing with different effectors would support the existence of a common

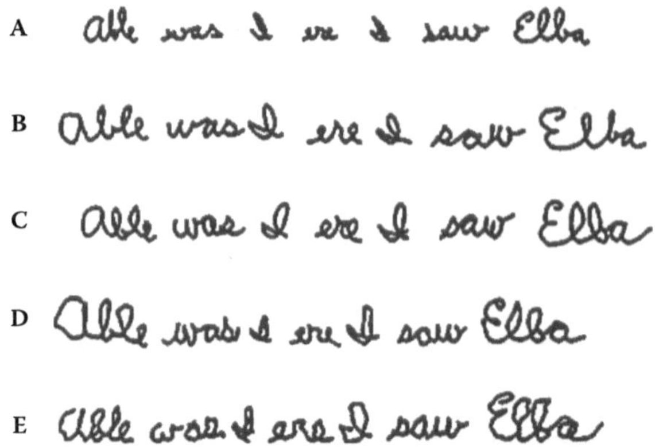

Figure 3.1 Example of motor equivalence for handwriting. Samples were written with the right (dominant) hand (A), with the right arm but with the wrist immobilized (B), with the left hand (C), with the pen gripped between the teeth (D), and with the pen attached to the foot (E). (From Raibert, M. H. 1977. Technical report, Artificial Intelligence Laboratory, MIT, AI-TR-439. With permission from MIT.)

network for writing behavior and validate the concept of motor equivalence. Subjects were instructed to sign their name or repetitively move the finger in an up-and-down zigzag manner in the air using either the hand or the toe. Various brain regions were imaged while subjects performed these "equivalent" tasks.

The investigators found that the same brain areas involved in finger zigzagging were also involved in name signing irrespective of whether the hand or toe was used. Interestingly, the cortical regions active during these tasks were located within the parietal sensorimotor areas that map anatomically to the handwriting areas described in Chapter 1. Figure 3.2 shows the activation patterns associated with finger and toe writing (signing) and nonwriting (zigzag) movements. These results provide anatomical support for motor equivalence as a control strategy for handwriting.

Motor equivalence is a product of a general principle of motor control that relies on a hierarchical network of modules responsible for storing specific parameters of the motor plan. The modular approach allows maximum flexibility and accommodation when faced with unpredictable environmental constraints. For example, in handwriting, the writer can maintain the planned action sequence and timing throughout changes in wrist angle, variation in writing surface, grip force, and pen orientation because the kinematic parameters for a given handwriting stroke are thought to be stored at a low level within the hierarchy (Teulings 1996). Whereas these muscle-independent parameters are likely hard-wired into the handwriting program, other parameters such as stroke size, starting position, muscle assignment,

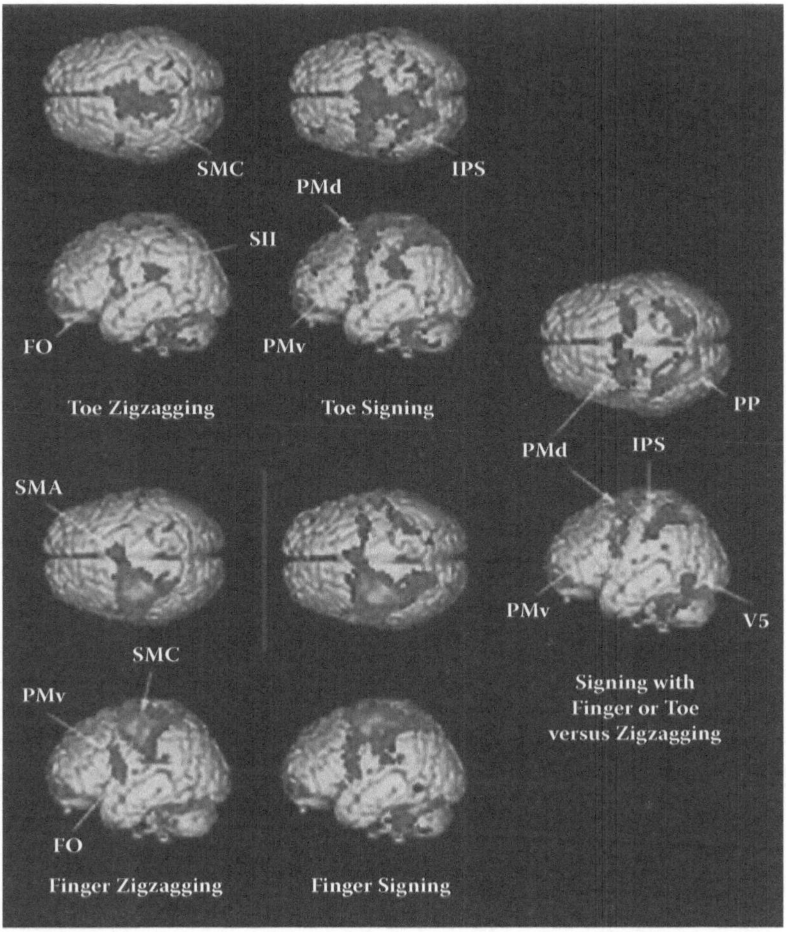

Figure 3.2 Brain activation showing overlapping cortical regions associated with repetitive finger and toe writing (signing) and nonwriting (zigzag) movements. Findings provide anatomical support for motor equivalence as a control strategy for handwriting. (From Rijntjes, M. et al. 1999. *Journal of Neuroscience* 19:8043–8048. With permission.)

and pen movement velocity are likely stored in higher level modules (van Galen and Teulings 1983) and can be set to meet specific task demands and environmental constraints.

Hierarchical models of handwriting generally consist of a few lower level modules (for kinematic storage and retrieval) and many higher level modules (for adapting to task demands). Van Galen et al. 1996 and van Galen 1991 proposed that the lower level modules are organized serially—that is, information is transferred from one submodule to another in sequence—whereas higher level modules are organized in parallel, permitting the simultaneous exchange of information from multiple domains. Under this scheme, the more dependent the writer is on higher level processing and programming of information, the more likely it is that there will be errors in the final action sequence. As processing demands increase, so do errors such as movement delays. For example, while stroke duration may be programmed and retrieved from a low-level module, initiation time and total writing duration are subject to parallel influence from higher level modules (van Galen 1986).

The concept that higher level properties of handwriting movement are more vulnerable to production errors than lower level (invariant) properties has implications for study for handwriting authentication. Invariant features of the motor program are thought to be unique to an individual. Within the context of a hierarchical structure, lower level invariant parameters are impervious to external influence or mechanical constraints. An individual should be able to reproduce a handwritten sample under a variety of cognitive and mechanical loads without discernable alteration to the lower level parameters. Conversely, with increased dependence upon information processing (such as when concentrating on the accuracy of a forged signature), it is likely that the reparameterization by higher level modular input will disrupt the execution of lower level parameters. As noted earlier in the general discussion of models, models enable the formulation of a testable hypothesis. One testable hypothesis derived from a hierarchical model of handwriting would predict that, in the face of increasing higher level modular demands (e.g., attempting to forge a signature), the resultant output should be deficient along several low-level kinematic parameters.

Cost Minimization Models

Several computational models have been proposed to enhance our understanding of complex goal-directed movement. Computational models were developed to explain how the nervous system governs specific kinematic properties of handwriting movement. Two popular computational models are the cost minimization models (Viviani and Flash 1995; Engelbrecht 2001) and the delta lognormal model (Plamondon and Guerfali 1998a; Plamondon and Djioua 2006; Djioua and Plamondon 2009).

Cost minimization models are based on the principle of optimal control (Flash and Hogan 1985). The kinematic profiles for rapid, simple, point-to-point arm movements are surprisingly stereotypical. Velocity profiles for movements varying in distance and duration are consistently bell shaped and symmetrical within and between individuals (Abend, Bizzi, and Morasso 1982; Atkeson and Hollerbach 1985; Morasso 1981; Miall and Haggard 1995). Experimental observations of the velocity profiles of a variety of horizontal plane hand and arm movements executed under different time constraints are consistent with cost minimization principles (Hollerbach and

Flash 1982; Soechting and Lacquaniti 1981; Abend et al. 1982; Morasso 1981). Engelbrecht (2001) proposed that this behavioral uniformity stems from the motor system preferring certain movement trajectories to others because of their efficiency.

Several influential models of motor control are based on the theory that movement parameterization is determined based on energy costs or error reduction. These are referred to collectively as cost minimization models (Hogan 1984; Flash and Hogan 1985; Viviani and Flash 1995, among others) and include the minimum jerk model, isochrony, and the two-thirds power law.

Minimum Jerk

Prior to becoming part of a hard-wired motor program, muscle action sequences are initially built upon a hierarchy of undetermined movement trajectories and torques that are eventually transformed to an appropriate movement sequence. One principle of cost minimization holds that these movement trajectories are selected such that the time integral of the squared magnitude of jerk (the third time derivative of position) is minimized (Hogan 1984; Flash and Hogan 1985). Thus, for a given movement, the trajectory is selected so that changes in acceleration are kept to a minimum. This is known as minimum jerk.

The cost savings in terms of energy and time realized by selecting a movement trajectory having minimum jerk thus free up the motor program from having to account for other undetermined properties of the movement task, such as the Cartesian space within which the movement is executed or movement duration. For a simple movement trajectory, the motor program need not concern itself with the constraints of the workspace (in two or three dimensions) or the time constraints imposed by the task. The movement trajectory from point A to point B is selected to reduce acceleration changes (or jerk).

When initially proposed by Flash, Hogan, and others in the early 1980s, the minimum jerk principle accounted for the majority of the kinematic observations available at the time. However, since then, a number of studies have identified inconsistencies between these principles and observation. For example, Weigner and Wierzbicka (1992) noticed that for horizontal-plane forearm movements having one degree of freedom (single joint movement), the ratio of peak to average velocity varied as a function of movement duration. That is, the velocity profiles were not symmetrical. Rather, for faster movements the velocity profile tended to be skewed to the left while slower movements had right-skewed velocity profiles (Moore and Marteniuk 1986; Weigner and Wierzbicka 1992). Thus, the prediction that the velocity of the movement trajectory is time independent may not hold for all types of movements.

Restricting movement to a single joint may constrain the motor program such that cost minimization is less optimal than for multijoint movements.

The minimum jerk principle does not necessarily explain all forms of movement, nor was it ever intended to. As a fundamental principle developed to account for why some movement trajectories are selected over others during simple horizontal-plane arm movements, the minimum jerk principle offers insight into the nature of the motor program and predicts observations when these programs fail.

Within the family of cost minimization principles, at least two alternatives to minimum jerk have been proposed. These include minimum torque (Uno, Kawato, and Suzuki 1989; Engelbrecht 1997) and minimum discomfort, which refers to reducing discomfort associated with endpoint posture (Cruse 1986; Rosenbaum et al. 1993). While attractive in that they have been shown to predict movement trajectories relatively independent of movement duration and have been experimentally observed in multijoint movements, neither of these fully accounts for the range of possible movement trajectories available during a movement action sequence.

Isochrony Principle

The isochrony principle states that "average velocity of point-to-point movements increases with the distance between the points and therefore that movement duration is only weakly dependent on movement extent" (Viviani and Flash 1995, p. 34)—in other words, equal durations for two trajectory components that differ in length. Early writings on the development and mechanisms of voluntary motor ability referenced the concept of isochrony long before it was formally studied as a minimization principle (Bryan 1892; Stetson and McDill 1923). The concept of isochrony suggests that some information stored in the motor program is constant, thus reducing the storage demands of the program. More formally, the idea of parametric constancy is reflected in Fitts' law (1954), which states that, under certain conditions, information output from the motor system is relatively constant. Fitts' law implies that movement time is an approximately linear function of distance.

Viviani and Terzoulo (1982) observed that the dynamical properties of movement such as velocity are largely determined by the movement trajectory. They argued that a valid test of the isochrony principle requires study of movements that involve reversals of direction and curvilinear, rather than linear, trajectories. They further demonstrated that, in handwriting, a strong relationship existed between the form of the trajectory and tangential velocity and that this relationship was evident within stroke segments of the handwriting movement. They observed that within each unit of action (i.e., stroke), the angle of the trajectory (and thus the length) was produced in roughly equal durations. The latter finding supports the idea that despite being overlearned, motor behavior handwriting is likely stored as individual concatenated segments.

An important finding to emerge from the Viviani and Terzoulo study was that the relationship between form or length of the trajectory and duration was observed for handwriting movements produced with or without visual feedback. This lends further support for the notion that the isochrony strategy is available within a central control process responsible for motor execution and therefore would not depend on sensory integration.

In a subsequent study, Viviani and Flash (1995) utilized a figure-drawing task to examine the durability of isochrony across various geometric constraints. Subjects were instructed to trace loop patterns composed of closed mathematical curves (asymmetric lemniscates), cloverleaves, and oblate limaçons. Using these templates, absolute and relative sizes of the loops could be specified by mathematical constants. Three healthy right-handed male subjects participated. Their results were consistent with the isochrony principle in that the reductions in loop perimeter (from 75 to 60 cm) spontaneously led to a reduction in average velocity so that cycle duration remained constant. Evidence of velocity scaling was present within each movement cycle of the overall multiloop pattern, thus demonstrating a local and global form of isochrony. Recall also that van Galen and Weber (1998) reported that while writers adjusted vertical stroke size to accommodate the spatial manipulation of writing line length, stroke duration was unchanged by this manipulation—another example of adherence to the isochrony principle.

These and other observations from the 1980s and 1990s have shown that the isochrony principle holds for almost any type of movement (Viviani 1986; Viviani and McCollum 1983; Viviani and Schneider 1991). Temporal isochrony is therefore one strategy available to the motor program for satisfying the problem of global optimization.

Further support for the isochrony principle in handwriting comes from studies of overlapping figure eights (Lacquaniti, Terzuolo, and Viviani 1983; Rosenbaum 2010). By extracting the angle of pen movement and the average angular velocity of each curve of the figure, one can plot the change in angular velocity as a function of time. When performed naturally, each loop of the figure eight is drawn with a constant angular velocity. More importantly, the total time to draw each loop is also equal. Producing equal angles in equal times conforms to the isogeny principle or, when referring only to the temporal component, isochrony (Viviani and Terzoulo 1982).

This observation has at least two implications for the theory of handwriting motor control. First, complex continuous curve writing is likely segmented into components each of which can be characterized by its own kinematic properties, such as angle, angular velocity, time, and length. This suggests that, in handwriting, individual letters are likely the manifestation of concatenated segments. Stroke-based models of handwriting incorporate the notion of segment (or stroke) concatenation as a strategy to overcome storage capacity limits (Bullock, Grossberg, and Mannes 1993; Rhodes et al. 2004). Second, the isochrony principle suggests that the fundamental control parameters in the execution of handwriting are likely to include time (i.e., stroke duration).

As a kinematic model of motor control, isochrony can be verified from just two parameters of movement: the length and duration of the segment or curve. As we will demonstrate in Chapter 8, adherence to the principle of isochrony can be evaluated directly from handwriting movements to test any number of hypotheses pertaining to differences between genuine and forged signatures.

Two-Thirds Power Law

Previous discussion of minimum jerk and the isochrony principle suggests that human motor control is organized and executed under maximum efficiency. It should be apparent that these principles of minimization dramatically reduce the computational burden of the motor program. The selection of a particular trajectory (to ensure minimum jerk) and increasing angular velocity to maintain constant movement duration are just two ways in which the motor program can simplify the demands of complex motor control. Yet, a third computational strategy is available to the motor program to ensure efficient and reliable motor control: the two-thirds power law.

The two-thirds power law describes the lawful relationship between angular velocity and curvature of movement (Lacquaniti et al. 1983, 1984). Derived from research on drawing movements, the two-thirds power law indicates that as the arc of the curvature of arm movement becomes more acute, the angular velocity increases. The increase in angular velocity is not linear, but rather curvilinear and can be expressed by the following equation:

$$A(t) = kC^{2/3}(t)$$

where
A denotes angular velocity at time point (t)
k is an empirical constant
C is curvature at time point (t)

The power law is an organizational principle that accounts for observed constancy in motor output produced by different effectors over different movement speeds and amplitudes (Vinter and Mounoud 1991; Viviani and Terzoulo 1980). Along with the other minimization principles, the power law increases the likelihood that the motor system will produce the desired goal under different spatial and temporal conditions regardless of the effector used (a critical requirement for handwriting).

The underlying mechanism accounting for this relationship is thought to stem from the coupling of two independent biological oscillators (Lacquaniti et al. 1984). Since a biological oscillator is considered sinusoidal with a specific phase (angle) and amplitude, combining multiple oscillators each with a unique phase and amplitude parameter can produce output curves of varying length and curvature (Hollerbach 1981; see later section on equilibrium point, or mass spring models). Thus, a simple curved motion of the arm or wrist could be accomplished by setting the phase (angular velocity) and the amplitude (curvature) of multiple oscillators.

It has been argued that while the two-thirds power law may be a good predictor of angular velocity of curved movement in most situations (Viviani 1986), the law may not apply to all forms of movement, particularly handwriting (Thomassen and Teulings 1985; Wann, Nimmo-Smith, and Wing 1988; Plamondon and Guerfali 1998b). The power law can be observed for a variety of handwriting forms ranging from ellipses to scribbling and for handwriting movements featuring multiple directional changes (Lacquaniti et al. 1983, 1984; Viviani and Flash 1995). However, growing evidence suggests that the two-thirds power law may not be as invariant a principle as initially conceived. For example, the power law relating angular velocity to curvature seems to vary with movement speed (Wann et al. 1988; Sailing and Phillips 2002), smoothness of movement, number of and particular joint involved in the movement (Saling and Phillips 2002, 2005), and movement size (Schaal and Sternad 2001). Saling and Phillips (2002) observed that the power law was stronger for faster than slower movements, for shoulder than for hand or wrist motion (Saling and Phillips 2002, 2005), and when handwriting movement involved fewer joints and smaller curvatures of motion (Phillips 2008).

Plamondon and Guerfali (1997, 1998a) observed that the power law could not predict angular velocity for single stroke handwriting movements with constant curvatures (i.e., where $C(t)$ at any given time point is equivalent to $C(t)$ at time point zero). For single stroke movements having constant curvature, the (instantaneous) angular velocity at time zero is proportional to the angular velocity at any other time point. Thus, the notion that the power law is derived from the coupling of different sinusoidal oscillators to account for differences in spatial (angular) and temporal constraints may not apply to single stroke handwriting movements derived from only one oscillator.

Furthermore, Plamondon and Guerfali (1998b) argued that the predictive utility of the power law requires that successive strokes (e.g., comprising a loop) must be out of phase by a constant factor (proportional to $\pi/4$) for a certain period of time. Natural handwriting does not follow this rule. Lastly, the two-thirds power law did not hold for some parts of the movement trajectory for nonoscillatory movements (Plamondon and Guerfali 1998b). These and other observations from natural and computer-simulated handwriting led to the development of an alternative to the two-thirds power law: namely, the delta lognormal model or kinematic theory (Plamondon and Guerfali 1998a; Plamondon 1998; Plamondon, Feng, and Woch 2003; Plamondon and Djioua 2006; Djioua and Plamondon 2010).

Kinematic Theory

Computational modeling of handwriting movements was the focus of a significant volume of research in the 1990s, culminating in what may be referred to as the kinematic theory of motor control (see Plamondon and Djioua, 2006, for review). Kinematic theory is derived from the notion that the neuromuscular system controlling rapid movement comprises subsystems coupled together to generate a desired velocity response. Each impulse response from this local system again converges onto a larger global network of systems. Inherent in the translation from local to global subsystem response are nonlinear time delays. These time delays can be mapped onto the desired velocity profile using a lognormal function. Thus, the equations derived from kinematic theory are closely related to the two-thirds power law as a parametric approach to modeling handwriting movement velocity (Djioua and Plamondon 2010).

The kinematic theory of motor control suggests that simple human movement is the manifestation of synergistic actions of agonist (e.g., an extensor) and antagonist (e.g., a flexor) muscle contractions leading to a measurable movement velocity of the effector (Plamondon 1993). Each muscle contraction has three properties that comprise a single impulse response: the delay of the impulse, the activation time of the impulse, and the response time. The resultant movement is modeled as the difference in the log function of these temporal parameters between the agonist and antagonist network—that is, a delta-lognormal model (Plamondon 1993, 1995; Plamondon and Guerfali 1998b). Using only three parameters (the starting point, the starting direction, and the curvature), Plamondon and Guerfali (1998a) accurately characterized single movement strokes having a lognormal velocity profile. The delta-lognormal model describes the lognormal relationship between multiple agonist and antagonist muscle pairs active within a global network during two-dimensional movement for the generated movement velocity.

As noted before, a fundamental property of the minimization principles is the notion that movement curvature may be modeled as the coupling of independent biological oscillators (Hollerbach 1981; Lacquaniti et al. 1984). The two-thirds power law suggests that simple curved motion of the arm or wrist could be accurately modeled by knowing the phase (angular velocity) and amplitude (curvature) of multiple oscillators. An intrinsic feature of kinematic theory is the lognormal translation of local agonist and antagonist muscle impulse responses to a global network of impulse responses to achieve a desired velocity response. In this sense, each agonist–antagonist impulse response may be thought of as an oscillator.

The notion that muscle synergies act as oscillators has been a key principle in many computational models of motor control. We conclude this chapter with a discussion of the equilibrium point (or mass spring) model and its attempt to characterize handwriting based on a very limited set of muscle parameters.

Equilibrium Point Model

The most common oscillatory model is the equilibrium point (or mass spring) model, which was developed to model fluent movement trajectories. The equilibrium point model assumes that harmonic oscillations represent the most fundamental mode of action of viscoelastic biomechanical systems (Hollerbach 1981). In this scheme, when attached to a mass, muscles act like springs. Since springs oscillate when set in motion, as mass is applied differentially to each spring, the form of motion changes. The mass inside the body (e.g., the arm) is suspended by surrounding tissues, represented by damped springs. When the system of springs is put into motion, it begins to oscillate. If mass is intermittently applied in a diagonal direction, the motion assumes the shape of a curve while the shape of the curve can be modified by altering the vertical or horizontal stiffness. With only an occasional impulse delivered in the proper sequence at the proper time, the mass would continue to oscillate. Hollerbach (1981) and others (Wann et al. 1988; Rosenbaum et al. 1995) reasoned that the mass spring equilibrium point model potentially simplifies capacity of the handwriting motor program for encoding complex curves.

In robotic handwriting, simply changing the stiffness of a spring will result in a change in shape of a loop or letter. By varying the equilibrium point, the stiffness, and initial conditions of the mass spring model, vertical and horizontal sinusoids can be coupled to form a fluent handwriting trajectory. Figures 3.3 and 3.4 from Hollerbach (1981) demonstrate the effects of simply changing the stiffness of a two-dimensional spring model of robotic handwriting. Vertical (A) and horizontal (B) velocity components are generated by the mass spring model and the resultant writing trajectory. Vertical accelerations are produced in writing the word "hell." One word has twice the amplitude as the other. Note the temporal agreement despite different magnitudes of acceleration. Thus, the mass equilibrium point model used to generate robotically produced handwriting adheres to

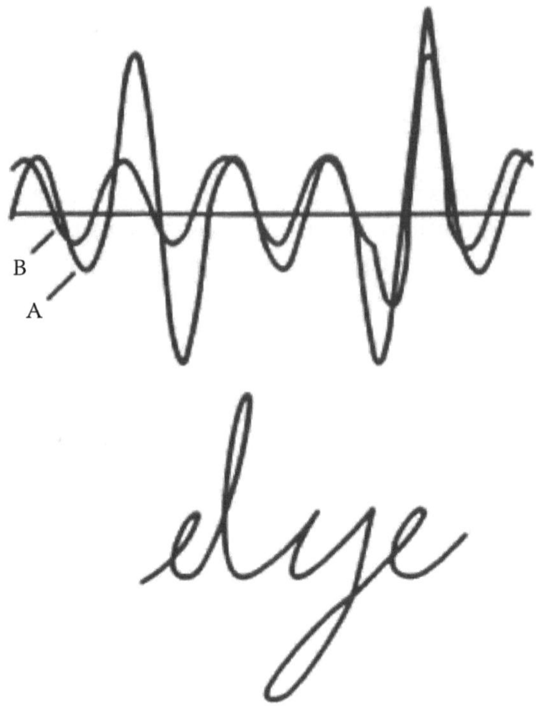

Figure 3.3 Effect of changing the stiffness of a two-dimensional spring model of robotic handwriting. Vertical (A) and horizontal (B) velocity components generated by the mass spring model and the resultant writing trajectory. (From Hollerbach, J. M. 1981. *Biological Cybernetics* 39:139–156. With permission.)

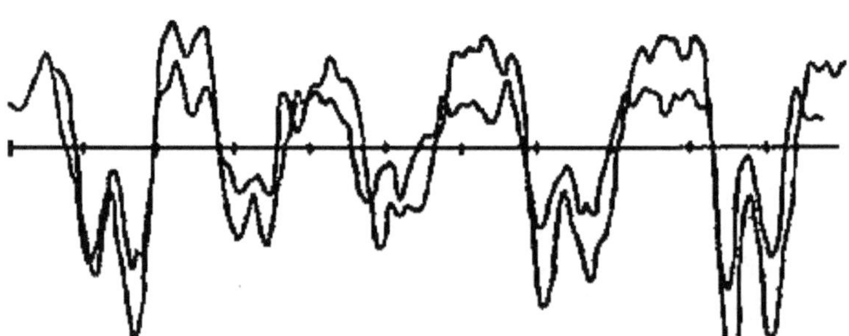

Figure 3.4 Vertical accelerations produced in writing the word "hell." One word has twice the amplitude as the other. Note the temporal agreement despite different magnitudes of acceleration. (From Hollerbach, J. M. 1981. *Biological Cybernetics* 39:139–156. With permission.)

the isochrony principle, suggesting that isochrony is a relatively low-level parameter contained within an effective handwriting program.

The equilibrium point model has significant limitations. First, normal handwriting does not have a consistent sinusoidal movement pattern (Teulings and Maarse 1984). Second, while the mass spring model is fairly accurate as a predictor of movement endpoint (Schmidt and Lee 1999), the model cannot account for starting position very well. In natural human handwriting, frequent stops and lifts of the pen inevitably lead to different restarting positions. At each restart, there could conceivably be a new equilibrium point, which may or may not be determined by the terminal equilibrium point. Thus, it is difficult to imagine how a handwriting motor program that relies upon viscoelastic biomechanical properties to determine movement endpoint can be efficient given the uncertainty and variability in starting position. Finally, while the equilibrium point model was developed and verified for movements involving only a single joint (e.g., elbow) in two-dimensional space, it is incapable of accounting for the coordinated multijoint movements (e.g., finger, wrist, and elbow) that accompany handwriting. While the equilibrium point calculation can successfully produce natural appearing handwriting by a mechanical arm, it falls short for multijoint human movements.

Summary

This chapter attempted to integrate an extensive body of research designed to elucidate the processes underlying complex coordinated handwriting movements. As a first step in understanding this process, scientists reduce its complexity into simpler manageable units or models. These manageable units can be visual, mathematical, or computational and can be used to develop testable hypotheses on the nature of motor behavior under various conditions. The models were developed not necessarily for the purpose of understanding handwriting, but rather for advancing our appreciation of how the human nervous system controls complex movement. While the translation of these general models of motor control to the specific problem of handwriting is often incomplete, there is sufficient overlap between handwriting movements and other complex movements, such as speech, typing, and other highly programmed movements, that these models have been successful in generating testable hypotheses and applied research in the areas of rehabilitation, movement disorders, and forensics.

The chapter addressed the controversy of whether handwriting stems from a motor program and, if so, what the program contains. A flexible motor program is often referred to as a generalized motor program and codes both variant and invariant movement parameters characterizing a given movement sequence. Generalized motor programs account for motor equivalence, or the ability to achieve the same movement outcome with different muscle groups. The most compelling evidence for the existence of a flexible and adaptive generalized motor program comes from empirical research on handwriting.

Nonetheless, consideration of handwriting as programmed motor behavior can be somewhat problematic. For one, it can be readily observed that handwriting is a serial motor behavior with individual letters making up words and words making up sentences in series. However, the existence of a motor program presumes that the movement parameters for handwriting are not stored as discrete instructions to specific muscles, but rather as a general spatial code representing the final motor output attainable under a variety of physical or environmental constraints. The ability of a writer to anticipate abrupt changes in the writing surface or writing instrument and evidence of motor equivalence provide strong support for handwriting as a highly flexible motor program.

We reviewed the current literature on relevant computational models developed to help elucidate the nature of the handwriting motor program, including hierarchical models, cost minimization models, and the equilibrium point model. Hierarchical models of motor control stem from cognitive approaches to understanding motor control. Such models account for both invariant (low level) and adaptive (high level) aspects of handwriting. Hierarchical models use a top-down structure, in which higher centers control or inhibit activity of the lower centers. In these models, control of motor output is decentralized with linear distributed control assigned to multiple levels within the hierarchy. As hierarchical models are compatible with a generalized motor program, they account for the observation of motor equivalence.

The concept that higher level properties of handwriting movement are more vulnerable to production errors than lower level (invariant) properties has implications for the study for handwriting authentication. Invariant features of the motor program are thought to be unique to an individual. Within the context of a hierarchical structure, lower level invariant parameters are impervious to external influence or mechanical constraints. An individual should be able to reproduce a handwritten sample under a variety of cognitive and mechanical loads without discernable alteration to the lower level parameters. Conversely, with increased dependence upon information processing (such as when concentrating on the accuracy of a forged signature), it is likely that the reparameterization by higher level modular input will disrupt the execution of lower level parameters.

Cost minimization models are computational models derived from observations with repetition and learning; human movement becomes programmed to follow optimal trajectories and time course to minimize cost in terms of energy and error. Researchers have demonstrated that handwriting movements subjected to various computational analyses are executed using stoke trajectories that are cost efficient. Efficient movement trajectories are those where jerk is minimized (i.e., reduced number of acceleration changes), movement time is constant despite changes in stroke length (the isochrony principle), and stroke velocity is determined by its curvature. These parametric rules simplify the demands of the motor program and allow greater flexibility and adaptation to environmental constraints.

Based on these three mathematical concepts, one would hypothesize that during production of a natural signature, the writer exhibits stroke parameters that adhere to a cost minimization principle, whereas in a forgery or disguised signature, the writer is likely to exhibit movement trajectories that are inefficient. Chapters 8 and 9 review findings from experiments designed to test this hypothesis.

Neurological Disease and Motor Control

4

Introduction

In this chapter, we provide an overview of common progressive neurological diseases and their effects on motor control in general and handwriting in particular. The aim of this chapter is to introduce the fundamental aspects of common neurological disease as the basis for an understanding of why and how handwriting changes in the presence of disease.

Pathological conditions that alter the neurotransmission within the sensorimotor areas of the neural axis can have profound effects on fine motor control of the hand. Neurological diseases often involve brain functions regulated by cortical, subcortical, brain stem, and peripheral processes. What distinguishes these progressive neurological diseases from the more episodic traumatic or cerebrovascular events is the insidious manner in which alterations in neurological function lead to motor impairment. Subtle changes in motor function over time signal the presence of a progressive disease process. This can be most apparent in Parkinson's disease (PD), in which gradual deterioration in handwriting is often the first sign that an individual may have PD.

Additionally, a host of cognitive diseases also impact motor function. The most common of these is dementia with Lewy bodies (DLB), a common form of Alzheimer's disease (AD). As a cognitive disorder, AD itself does not usually present as a movement disorder. When handwriting is impaired in AD the pattern reflects disruption of cognitive processes such as memory and sequencing. However in DLB, the motor dysfunction is characterized by psychomotor problems (disrupted timing and sequencing) along with classical parkinsonian motor features such as micrographia, slowness (bradykinesia), and, in some cases, tremor. Differential diagnosis is challenging because many PD patients develop dementia later in the course of their illness, blurring the distinction between PD and DLB.

We focus on the more common neurological conditions likely to present challenges to the forensic document examiner. While the quality and characteristics of an individual's signature are known to change gradually with age (see Chapter 6), diseases of the nervous system accelerate or alter this age-related transformation. In the following sections, we review the epidemiology, pathophysiology, and clinical characteristics of common neurological diseases likely to occur in an aging population, including PD and related parkinsonian disorders, essential tremor, ataxia, multiple sclerosis, Huntington's disease, lower motoneuron disease, and AD.

Parkinson's Disease

Parkinson's disease is a progressive disorder characterized by three cardinal motor signs: bradykinesia (slowness), rigidity (muscle stiffness), and tremor. The etiology in PD is known to involve cell loss within the substantia nigra pars compacta, the nuclear region that provides dopamine to the striatum. In 1817, James Parkinson first reported a syndrome consisting of tremor and postural instability affecting six of his patients. It was not until 1861 and 1862 that Jean-Martin Charcot (1825–1893) with Alfred Vulpian (1826–1887) added more symptoms to James Parkinson's clinical description, which subsequently confirmed his place in medical history by attaching the name Parkinson's disease to the syndrome.

The incidence of PD varies with age, ranging from approximately 0.02% for individuals between 50 and 60 years of age to 0.09% for those between 70 and 80 years of age. The lifetime incidence of PD (that is, the likelihood that an individual will be diagnosed with PD at some time in his or her life) is estimated at 1.5% (Bower et al. 1999). The onset of PD is gradual. The earliest motor signs are often indistinguishable from those associated with normal aging. Nonspecific complaints such as stiffness or slowness rarely prompt a visit to the local neurologist. While muscle stiffness is common among the elderly, as are general fatigue and motor slowing, other features of PD, such as tremor, shuffling gait, and loss of facial expression, are indicative of an abnormal aging process.

Its gradual onset and link to age have sparked debate as to whether PD is a pathological condition or simply accelerated aging such that if we lived long enough we would all develop parkinsonism in some form (Hindle 2010). Many of the age-related declines in motor function in humans are thought to be, at least in part, related to decline in central dopamine function. One of the most predominant findings in normal aging is the reduction in the width of the substantia nigra (Pujol, Junque, and Vendrell 1992). By the time motor symptoms appear, 70%–80% of dopaminergic neurons are already lost (Pearce 2008). Early physical signs generally precede the diagnosis of PD by 3–4 years (Morrish et al. 1996). PD is distinguished from other progressive neurological diseases by the presence of a cluster of clinical motor signs. These include akinesia (poverty of movement), bradykinesia, rigidity, tremor, stooped posture, and masked faces (loss of facial expression). Jankovic (2008) groups the four cardinal features of PD under the acronym TRAP (tremor, rigidity, akinesia, and postural instability).

The tremor in PD is a tremor at rest. Resting tremor is characterized by a 4–6 Hz (cycles per second) rhythmic movement of moderate amplitude of an extremity, usually the hand (Koller 1984; Findley, Gresty, and Halmagyi 1981). Parkinsonian tremor can be distinguished from other tremors, such as those associated with essential tremor or cerebellar disorders (see later discussion) by its frequency and amplitude characteristics. Essential tremor is characterized by a higher frequency oscillation (5–8 Hz) and lower amplitude, while cerebellar disease produces tremor with lower frequency oscillations (3–5 Hz) than in PD.

The term bradykinesia literally means slow movement. Bradykinesia is often used interchangeably with other terms to describe the range of mobility impairment in PD. In addition to slowness, arm and hand movements in PD may have reduced amplitudes or hypokinesia. Akinesia refers to the lack or reduction of spontaneous movement. In severe forms of akinesia, PD patients will freeze or come to a complete stop while walking. Micrographia is a common manifestation of parkinsonism encompassing all three features: akinesia, hypokinesia, and bradykinesia. Micrographia may appear as an early sign prior to the diagnosis of PD (Pearce 2008). Other secondary signs considered to be advanced manifestations of bradykinesia include impaired speech, postural instability, and festinating gait (shuffling). The core features appear asymmetric at the onset of the disease.

When applied to a clinical setting, rigidity refers to resistance to passive movement. In practice, rigidity is assessed by rotating a patient's arm or leg while the patient is at complete rest and sensing the resistance to this movement. An experienced clinician will distinguish resistance due to gravity and be able to judge whether spontaneous muscle activity is imposing force to resist free movement. In the laboratory setting, rigidity can be detected as increased stiffness (Caligiuri and Galasko 1992). Stiffness is defined as the ratio of rotational torque (or force) over displacement. In parkinsonian rigidity, increased muscle stiffness leads to an increase in resistance to passive movement. The Froment reinforcement maneuver (voluntary movement of the contralateral limb) will increase the stiffness in the test limb, making rigidity easier to detect in mild cases (Broussolle et al. 2007). Rigidity and bradykinesia often co-occur, rendering hand movements slow and laborious.

Progressive Supranuclear Palsy and Corticobasal Degeneration

Progressive supranuclear palsy (PSP) is a common extrapyramidal disorder characterized by postural instability and supranuclear gaze palsy. Recently published case reports have contrasted the clinical features of PSP with PD and concluded that, while there are similarities in the motor, cognitive, and behavioral features of the two conditions, the onsets of the clinical signs and relation to disease severity distinguish PSP from PD (Cordato et al. 2006). PSP also shares pathological characteristics and clinical features with corticobasal degeneration (CBD). However, unlike PSP, which manifests primarily as a subcortical movement disorder, the motor signs observed in CBD reflect higher cortical dysfunction (Soliveri, Piacentini, and Girotti 2005), enabling a reliable differential diagnosis.

For example, reporting on limb apraxia in CBD, Soliveri and colleagues (2005) allude to CBD patients' difficulty in performing individual finger movements and sequences of finger movements. The clinical presentation in PSP is consistent with neuropathological findings of increased tau protein in the neurons and glia throughout the brain stem, the substantia nigra pars compacta, subthalamic nucleus, and globus pallidus and, in CBD, the cortex, particularly the superior frontal gyrus (Arvanitakis and Wszolek 2001).

In a recent report comparing the motor characteristics of PSP and PD, Cordato et al. (2006) noted that while nearly 25% of their PD patients exhibited micrographia, the prevalence of micrographia in PSP was only 10%. However, 25% of the PSP patients produced handwriting samples that were described as laborious. Handwriting samples in PSP revealed several characteristics not observed in parkinsonian micrographia, such as untidiness, abnormal slanting, and illegibility. While the published literature on handwriting in PSP is not extensive, a clinical picture emerges that handwriting impairment in PSP is qualitatively different from that in PD. In some patients with PSP, handwriting impairment is likely to appear early in the course of their disease (Ahmed et al. 2008).

Essential Tremor

While motor signs such as slowness, muscle rigidity, and postural instability, are generally associated with a parkinsonian syndrome, tremor is more ubiquitous. Tremor can develop in people, regardless of age, for many reasons, including side effects of medications (to be discussed in Chapter 5), generalized anxiety, muscle fatigue, and other disease conditions. One common neurological condition in which tremor is a main feature is essential tremor (ET). The pathophysiology of ET is not fully understood; however, functional neuroimaging evidence has implicated abnormalities within the inferior olivary bodies, locus coeruleus, thalamus, and cerebellum (Bhidayasiri 2005).

Essential tremor is characterized as a postural tremor having multiple etiologies (Marsden, Obeso, and Rothwell 1983; Findley and Koller 1987; Jankovic 2002). The tremor frequency of ET ranges from 4 to 9 Hz, with a modal frequency of 6 Hz (Salisachs and Findley 1984). Essential tremor tends to be of relatively low amplitude and is characterized by a synchronous agonist/antagonist burst pattern (Findley et al. 1981). Subtypes of ET have been described (Louis, Ford, and Barnes 2000) and can be distinguished on the basis of age of onset, anatomic distribution, and rate of progression (Calzetti et al. 1987).

Despite these apparent physiological differences between ET and PD, ET is often misdiagnosed as PD (Koller 1984; Elble 2002; Bhidayasiri 2005). The classic resting tremor of PD may be observed in patients with ET and, conversely, postural tremor is common in PD (Findley et al. 1981). Jankovic (2002) recognized the importance of obtaining an accurate history of the tremor and presence of other motor signs suggestive of parkinsonism when differentiating ET from PD. However, in the absence of a reliable tremor history, clinicians must rely on observable characteristics such as tremor frequency and amplitude and how they change when shifting from rest to posture. Yet, frequency and amplitude information has been of little value because of the difficulty in obtaining reliable judgments of these attributes.

An alternative approach to tremor assessment has been to employ sensitive instrumentation such as accelerometry. Burne et al. (2002) used both accelerometry and electromyography to discriminate between ET and PD on the basis of tremor amplitudes, frequency, and muscle burst patterns under postural and resting conditions. They found that while no single variable correctly classified all patients, a three-factor model consisting of tremor frequency and two selected amplitude parameters obtained from the resting limb discriminated 86% of the PD and 95% of the ET patients.

Multiple System Atrophy

Multiple system atrophy (MSA) is a sporadic, progressive disorder affecting the basal ganglia, cerebellum, motoneurons, and autonomic function to varying degrees. Included under this class of disorders is olivopontocerebellar atrophy (OPCA), striatonigral degeneration (SND), and Shy–Drager syndrome. When the disease involves the basal ganglia, as in SND, the resultant motor manifestations resemble parkinsonism. As in idiopathic PD, SND leads to loss of dopamine neurotransmission to the striatum and subsequent dysregulation of striatopallidal GABAergic outflow. Behaviorally, SND is difficult to distinguish from PD based on motor presentation alone.

When MSA involves the cerebellum, as in OPCA, the principal clinical motor manifestations are ataxia and tremor. Ataxia is a term used to refer to loss of synergistic movement. In cerebellar damage, coordinated, multijoint movements requiring precise timing are no longer executed in a fluid manner. Rather, complex movements are decomposed into individual units creating errors in timing, distance, and trajectory (Brooks 1986).

As a further complication, degeneration of pathways to and from the cerebellum produced an intention or action tremor. Unlike parkinsonian resting tremor, or the postural tremor of ET, cerebellar action tremor appears in the hand and arm during reaching movements. Action tremor has a frequency range of 3–5 Hz and increases in amplitude with increased need for precise aiming or as the hand moves closer to the target. Simply holding a pen may not trigger action tremor in patients with cerebellar disease; however, upon initiating hand movement, as the pen tip reaches closer to the paper, action tremor develops and increases in amplitude. Action tremor generally persists throughout the purposive movement.

Multiple Sclerosis

Multiple sclerosis (MS) is considered an autoimmune inflammatory disease likely triggered by viral infection (Lucchinetti 2008). The disease is marked by the progressive demyelination of nerves throughout the cortex, particularly the frontal lobes. Myelin is the whitish sheath made up of lipids and proteins that surround axons within the central nervous system (brain and spinal cord)—thus the term *white matter.* Myelin serves to increase the rate of electrochemical transmission through the nerve by allowing impulses to "jump" between myelinated junctions along the nerve. As the myelin degenerates, the axons can no longer effectively conduct signals, leading to a host of behavioral and functional consequences.

MS affects about 1 in 500 persons worldwide (Rosati 2001) and is often diagnosed early in adult life. Nearly 70% of patients manifest motor and cognitive symptoms between ages 21 and 40. The disease rarely occurs prior to age 10 or after age 60. Females are two to three times more likely to develop MS than males. The incidence of MS in first-degree relatives is 7–20 times higher than in the general population, suggesting the influence of genetic factors on the disease. Symptoms of MS usually appear in episodic acute periods of worsening or relapse throughout the gradually progressive deterioration of neurological function (Lublin and Reingold 1996). In MS, relapses are often unpredictable, occurring without warning and without obvious inciting factors with a rate rarely above 1.5 per year (Rosati 2001).

Clinical characteristics of MS include changes in sensation such as loss of sensitivity or tingling, pricking, or numbness; muscle weakness; muscle spasms or jerks; and difficulty in mobility, coordination, and balance. Tremor is relatively common in MS and is characterized by a slow, high-amplitude tremor and worsens at the end of an intended movement, most noticeably in the hands. Thus, tremor in MS is similar to the action tremor associated with cerebellar disease, can affect handwriting, and likely worsens with increased writing time. It is estimated that 75% of MS patients exhibit tremor that affects handwriting (Wellingham-Jones 1991).

Huntington's Disease

Huntington's disease (HD) is an inherited, progressively disabling disorder that causes problems with behavioral control, cognition, and motor function. While not the first to describe the disease, George Huntington published the most thorough description of the condition in 1872. The classic sign of HD is a dance-like involuntary movement, called chorea. The movement disorder in HD is a hyperkinetic disorder characterized by dyskinesia, the random involuntary movements of the limbs and trunk. HD is associated with progressive degeneration of cells within the basal ganglia, primarily within the caudate and putamen (collectively referred to as the striatum). As you may recall from Chapter 1, the striatum is the primary output nucleus of the basal ganglia subserving key motor functions. The degenerative changes in HD primarily affect the medium-sized "spiny" neurons located within the striatum. These are the neurons that project GABA (an inhibitory neurotransmitter) to the globus pallidus.

Consequently, loss of GABAergic inhibition within the basal ganglia motor circuit pathways leads to excessive, disorganized, hyperkinetic movement patterns. However, the progressive neural degeneration in HD is not limited to the basal ganglia. As the disease progresses, there is marked degeneration within the frontal and temporal lobes of the cortex, which accounts for the significant cognitive decline and dementia in HD.

Due to its heritability, the epidemiology of HD is well understood (Conneally 1984). The prevalence (the frequency of all current cases within a specific population) of HD is estimated to be only 50–90 per million; however, in certain regions of the world (e.g., Tasmania and western Venezuela), the prevalence is much higher. As with all autosomal dominant diseases, the HD gene (located on one of the nonsex chromosomes) is always expressed, even if only one copy is present. Thus, the offspring of an HD parent has a 50% chance of receiving and expressing this genetic condition. Unfortunately, symptoms do not appear until well after the childbearing years, so it is not always known whether the parent is a carrier until after he or she has offspring.

A genetic test is available for confirmation of the clinical diagnosis. In this test, a small blood sample is taken, and DNA from it is analyzed to determine the CAG repeat number. A person with a repeat number of 30 or below will not develop HD. A person with a repeat number between 35 and 40 has a high likelihood of developing the disease sometime within his or her normal life span. A person with a very high number of repeats (70 or above) is likely to develop the juvenile-onset form.

The symptoms of HD fall into three categories: motor, behavioral, and cognitive. The severity and rate of progression of each type of symptom can vary from person to person. Early motor symptoms include restlessness, twitching, and a desire to move about. Handwriting may become less controlled, and coordination may decline. Later symptoms include dystonia, or sustained abnormal postures, including facial grimaces, a twisted neck, or an arched back; chorea, in which involuntary jerking, twisting, or writhing motions become pronounced; slowness of voluntary movements; inability to regulate the speed or force of movements; inability to initiate movement and slowed reactions; difficulty speaking and swallowing due to involvement of the throat muscles; localized or generalized weakness and impaired balance ability; and eventually muscle rigidity. Personality and behavioral changes include depression, irritability, anxiety, and apathy. While the handwriting in HD will be discussed in greater detail in Chapter 10, it is interesting to note that the primary features that characterize the handwriting impairment in HD are the presence of movement interruptions and excessive number of velocity and acceleration reversals within stroke and excessive variability between strokes (Phillips et al. 1994, 1995; Iwasaki et al. 1999). As suggested by the progressive nature of the basal ganglia pathology, patients with advanced forms of HD may exhibit parkinsonian micrographia. Overlap in the pathophysiology of PD and late-stage HD and consequent manifestations in handwriting present challenges to both the treating clinician and the document examiner.

Lower Motoneuron Disease

Lower motoneuron diseases are a group of progressive neurological conditions that destroy motor neurons—the cells that control essential voluntary muscle activity. Destruction of the lower motor neurons and subsequent muscle deinnervation lead to loss of strength, atrophy (wasting away of muscle mass), and involuntary muscle twitching (called fasciculations). The causes of sporadic, or noninherited, motoneuron disease are not known, but environmental, toxic, or viral factors may be implicated. Sporadic cases may be triggered by cancer or prolonged exposure to toxic drugs or environmental toxins. It is also likely in some cases that motoneuron disease may be an autoimmune reaction to viral infection (e.g., HIV).

Among the many motoneuron diseases, amyotrophic lateral sclerosis (ALS, or Lou Gehrig's disease) is the most common. ALS is a progressive, ultimately fatal disorder that eventually disrupts signals from the brain to all voluntary muscles. Unlike the previously discussed progressive neurological diseases, ALS affects motor function exclusively. Behavioral or cognitive changes are rare and when present (e.g., depression) are considered secondary to the primary motor impairment. The earliest motor symptoms usually appear in the arms and hands, legs, or muscles of mastication.

Prevalence estimates indicate that approximately 30,000 people in the United States have ALS with an incidence of 5,000 new cases each year. ALS is more prevalent in men than women; first symptoms appear between the ages of 40 and 60. Most cases of ALS are sporadic—the cause is unknown and there is no known genetic association for ALS. However, there is a familial form of ALS in adults, which often results from mutation of the superoxide dismutase gene located on chromosome 21 and a rare juvenile-onset form of ALS is genetically transmitted. The life expectancy from the onset of symptoms is between 3 and 5 years; however, about 10% of affected individuals survive for 10 or more years. Nonetheless, when discussing survival following onset of ALS, one cannot resist the remarkable story of Stephen Hawking, theoretical physicist and former Lucasian Professor of Mathematics at the University of Cambridge, who developed ALS over 48 years ago at age 21.

As noted, the primary motor signs in ALS include muscle weakness, atrophy, and fasciculations. It should be obvious that muscle weakness can have profound effects on handwriting, including slowness, inability to maintain handgrip, reduced pen pressure, and rapid fatigue leading to the inability to maintain the appropriate hand and wrist posture necessary to produce even a single handwritten letter. Systematic handwriting research has not been conducted in lower motoneuron disease per se; however, there have been a few studies on handwriting of individuals with muscle fatigue (Provins and Magliaro 1989; Poulin 1999; Harralson, Teulings, and Farley 2009). In general, this research shows that muscle fatigue can deteriorate handwriting by prolonging the temporal components of stroke production and increasing between stroke variability in such kinematic features as stroke amplitude and speed.

Harralson conducted a systematic study of the effects of inducing fatigue among patients with PD, essential tremor, and healthy comparison subjects. Subjects were asked to perform 24 handwriting tasks including signatures, sentences, and spirals in succession. It was assumed that fatigue would become a factor over the 24 trials. Handwriting samples were digitized and their kinematic features analyzed. Most subjects exhibited an increase in stroke size over time. Those with tremor developed worsening of tremor, which appeared in the handwriting trace. While these findings on induced fatigue cannot generalize to patients with pathological fatigue due to muscle atrophy, as in ALS, they do demonstrate what can be expected in the handwriting of patients with lower motoneuron disease.

Alzheimer's Disease and Dementia with Lewy Bodies

No chapter on the implications of progressive neurological disease on forensic document examination would be complete without special attention to dementia, particularly Alzheimer's disease. AD is a prevalent dementing disorder characterized by progressive loss of memory followed by gradual deterioration of judgment, reasoning ability, verbal fluency, and other cognitive skills. Late in the disease progression, AD patients develop behavioral problems such as wandering, hostility, and regressive behavior. Symptoms develop slowly and progress over time, eventually becoming severe enough to interfere with simple daily tasks such as feeding, dressing, and hygiene.

Estimates of the incidence (number of new cases per year) for dementia range from 10 to 15 new cases per thousand per year for all dementias and from 5 to 8 for AD (Bermejo-Pareja et al. 2008; Di Carlo et al. 2001). Thus, AD accounts for approximately 50%–60% of all dementing illnesses. AD has an average age of onset of 65 years old. However, in 5% of the cases, onset of symptoms appears at a much younger age, typically between 40 and 50 years old. The risk of developing AD increases dramatically with age. For example, every 5 years after the age of 65, the risk of acquiring the disease approximately doubles, increasing from 3 to as much as 69 per 1,000 per year beyond 85 years of age (Bermejo-Pareja et al. 2008; Di Carlo et al. 2001). Women have a higher risk of developing AD than men, particularly in the population older than 85 (Andersen et al. 1999; Hebert et al. 2003).

Accurate diagnosis of AD cannot be confirmed except through autopsy. Individuals who present with the cognitive features of AD are usually given the diagnosis of *probable* AD. Nonetheless, because correlational studies strongly support the link between autopsy-confirmed pathological changes in AD and clinical presentation, the term "probable" has been dropped from most clinicians' vocabularies. The pathology of AD was first described by Alios Alzheimer in 1906. The two hallmark features of AD observed at autopsy are amyloid plaques and neurofibrillary tangles. Plaques are extracellular deposits of abnormally processed amyloid precursor protein, and tangles are intracellular accumulations of the cytoskeletal protein tau. It is now recognized that the development of plaques and tangles may not be responsible for the early biochemical changes in the AD brain. Rather, other processes such as inflammation, disruptions of cell signaling pathways, and cardiovascular factors appear to play important roles early in the disease.

A significant portion of individuals with AD develop parkinsonian motor features. When this occurs, the provisional diagnosis of dementia with Lewy bodies (DLB) is made. Lewy body formation in the brain is pathognomonic of Parkinson's disease. Thus, DLB patients exhibit dementia with all the classic parkinsonian motor features. Epidemiological studies suggest that approximately 20% of patients meeting clinical criteria for AD at autopsy have neocortical Lewy bodies in addition to the classical AD lesions of neuritic plaques and neurofibrillary tangles (Hansen and Galasko 1992; Samuel et al. 1997). Lewy bodies are present throughout the parkinsonian brain, including the substantia nigra, locus coeruleus, basal nucleus of Meynert, amygdala, and hippocampus (Forno 1996; Dickson 2001; Jellinger and Mizuno 2003) and have been linked to progression and staging of PD (Braak et al. 2004). DLB constitutes the second largest group of dementia patients after "pure" AD (Perry, Irving, and Thomlinson 1990).

Clinically, DLB overlaps with both AD and PD, making differential diagnosis a challenge. The Consortium on Dementia with Lewy Bodies (McKeith et al. 1996) identified cognitive impairment progressing to dementia as the central feature of DLB. Additional features of DLB include specific cognitive impairments such as deficits in attention, problem solving, and visuospatial function. Fluctuating cognitive function, persistent visual hallucinations, and spontaneous extrapyramidal motor features are considered core features of DLB and have been shown to discriminate DLB from AD and other dementias (McKeith et al. 1996; Geser et al. 2005; Tiraboschi et al. 2006). DLB patients are more likely to develop intolerance to antipsychotic medications (McKeith et al. 1992) and their illness tends to progress more quickly than AD (Hanyu et al. 2009). Current practice is to restrict the diagnosis of DLB only to patients presenting extrapyramidal motor signs and concurrent dementia having onset within 1 year of the motor signs (Geser et al. 2005).

Handwriting movements of most AD patients remain relatively preserved throughout their lives, normal aging effects notwithstanding. Exceptions are when fine motor skills are affected by psychomotor processes, such as timing and sequencing and in DLB. While there have been numerous published works characterizing handwriting across the spectrum of neuromotor disease (briefly cited in this chapter and in greater detail in Chapter 10), there has been little effort to characterize handwriting in AD. Behrendt (1984) was among the first to call attention to this problem from the perspective of the forensic document examiner, noting that "the writing of these people will many times contain the normal indications of the aged, senile writers, for example, omission of letters, repetition of letters, and improper connection of words, yet show very little loss in writing skill as would normally be expected" (p. 86).

AD patients eventually lose the ability to sign their name on command (reflecting a cognitive rather than neuromotor process); however, with sufficient prompting and use of a model, they can produce an effective signature. This, of course, presents special problems for the forensic document examiner. Behrendt noted that one should not expect any decline in handwriting skill as severity of dementia increases. This astute observation has been confirmed by modern systematic studies (e.g., Schröter et al. 2003).

Schröter and colleagues (2003) evaluated handwriting movements in patients with AD and mild cognitive impairment (MCI) compared to those of healthy subjects to test whether these groups differed systematically on measures of handwriting kinematics and whether handwriting dysfunction can be used to differentiate patients with mild forms of cognitive impairment from those with AD. Subjects were instructed to draw concentric superimposed circles as fast and fluently as possible with and without a distraction task. Measures of handwriting speed (frequency or number of circles per second, velocity, and variability in velocity between strokes) and smoothness (changes in velocity direction) were extracted.

With regard to the differences between subjects with MCI and AD, the authors reported that AD patients exhibited significantly greater variability in velocity than MCI and healthy subjects; however, no differences were found in movement speed or frequency. Age, but not dementia severity, was correlated with handwriting kinematics in AD. These findings suggest that in the absence of overt motor impairment such as parkinsonism, sensitive measures of handwriting movements can reveal subtle impairments in AD.

While there is a clear need for more research, handwriting in AD appears to be characterized by the preservation of kinematic features such as speed, stroke duration, and size (adjusted for age), with increased variability and loss of smoothness and fine control. Debate remains as to whether the decline in handwriting in AD reflects the pathological change in frontal cortical integrity, giving rise to cognitive and psychomotor deficits (Slavin et al. 1999), or pathological change in subcortical basal ganglia integrity, giving rise to parkinsonian features as in DLB. It is likely that both processes are involved.

Summary

The purpose of this chapter was to survey common neurological diseases and their consequent impact on motor control. One of the major sources of variation in handwriting over time, particularly among older writers, is the effect of progressive neurological disease. Estimating how a given neurological condition affects handwriting requires an understanding of the relationships between normal (Chapters 1 and 2) and pathological (this chapter) neuroanatomy, neurochemistry, and motor function.

The study of neurological disease has expanded our understanding of central nervous system control of motor functions. With the exception of dementia, progressive neurological diseases generally affect deep brain centers involved in multiple aspects of motor control. Diseases such as Parkinson's disease, progressive supranuclear palsy, and Huntington's disease disrupt neurotransmission to important basal ganglia nuclei regulating of motor control. Interruption of basal ganglia circuits can lead to restricted (as in PD) or excessive (as in HD) movements. Diseases affecting lower motor neurons (such as ALS) produce muscle weakness. Hand movements in lower motoneuron disease are slow and executed with reduced force. Dementing illnesses such as Alzheimer's disease may not involve degeneration of cortical or subcortical motor areas. The motor control deficits exhibited by AD patients are characterized by higher level psychomotor abnormalities, which may or may not involve handwriting. However, with sufficient cognitive impairment, complex movements will show degradation. Some patients develop an interesting subtype of AD known as dementia with Lewy bodies. These patients exhibit the same cognitive and behavioral declines as in typical AD with the additional problem of parkinsonism. Hand movements in patients with DLB resemble those of PD patients.

Because of the complex integration between the cortex and subcortical, cerebellar, and brain stem nuclei, disease processes affecting one site can influence neurotransmission throughout the circuit. Thus, attempts to differentiate among various neurological conditions based on assessment of motor function for the purpose of diagnosis can be futile. Nonetheless, an examiner with a limited appreciation of specific patterns of motor dysfunction associated with neurological diseases can begin to identify potential sources of variability in handwriting.

Psychotropic Medications

Effects on Motor Control

5

Introduction

This chapter provides an overview of psychotropic medications and their effects on motor control and handwriting. We chose these medications as the focus because of their widespread use across multiple demographic groups in modern society. Psychotropic drugs include a broad range of pharmacological agents, including antidepressants, anxiolytics, antipsychotics, and mood stabilizers.

The decision by a physician to prescribe a given psychotropic medication is not necessarily driven by diagnostic criteria, but rather by symptoms and clinical course as there is overlap in symptoms and treatment response across multiple diagnostic classifications. For example, it is not uncommon to prescribe an antipsychotic to manage symptoms associated with depression, sleep disorders, anxiety, and dementia. The basis for this practice is that the available psychotropic medications act on a limited number of neurotransmitters that mediate diverse and complex behaviors and emotions. As these neurotransmitters subserve both emotional and motor functions, the net effects of many pharmacological interventions are not always desirable. For example, a drug that blocks dopamine neurotransmission can reduce psychosis but can also alter motor functions, motivation, and arousal. Drugs that target the serotonin pathways not only alter mood regulation and temperament, but also, because serotonin modulates dopamine, affect all of the dopamine-mediated behaviors (such as motor function and motivation).

Psychotropic medications target specific regions of the brain that regulate mood and emotion and they alter behavior in two ways. The molecules of common psychotropic drugs bind to receptors and block the transmission of such neurotransmitters as dopamine, GABA, or serotonin to travel from one neuron to another within a circuit, thereby reducing availability of the neurotransmitter. Conversely, the molecules can bind to receptors that permit the reabsorption of excess neurotransmitter within the synaptic junction, thus decreasing the turnover and increasing the availability of neurotransmitter.

The primary sites of action for most psychotropic drugs are within the limbic and mesolimbic system (subcortical brain structures) as well as the motor regions of the basal ganglia. This is important because of the anatomical, neurochemical, and functional overlap within the limbic system (for mood regulation) and those of the basal ganglia (for motor regulation). As such, psychotropic medications impact neurotransmission to impart change within the emotional and motor circuits of the brain. Unfortunately, as will be discussed in greater detail later, the therapeutic effects of psychotropic medications on mood and emotion are inevitably accompanied by countertherapeutic effects on the motor system.

Given the ubiquitous accessibility of psychotropic medications today, particularly in the aging population, it is important that the forensic document examiner gain an appreciation of the potential influence of these common medications on handwriting. In the following sections, we present an overview of the different classes and types of psychotropic drugs, mechanisms of action and clinical indication, and specific effects on the motor system.

An Overview of Psychotropic Medications

Table 5.1 lists the commonly prescribed psychotropic medications, mechanisms of action, indications, and common motor side effects. This table includes only medications that are FDA approved for specific indications in the United States.

For many patients, the therapeutic benefits of psychotropic medications are outweighed by the countertherapeutic motor side effects. Antipsychotics are not the only psychotropic agents that can produce motor problems such as parkinsonism and dyskinesia (discussed in greater length later), both of which can affect handwriting. Studies have also shown that some classes of antidepressants, particularly selective serotonin reuptake inhibitors (SSRIs) such as fluoxetine (Prozac), can produce movement abnormalities in vulnerable patients. For example, Leo (1996) reported that 14% of the cases treated with SSRIs developed parkinsonism (including bradykinesia, rigidity, and tremor) and 11% developed tardive dyskinesia. Gerber and Lynd (1998) reviewed 127 published reports of SSRI-induced movement disorders and found SSRI-induced parkinsonism in as many as 19.7% of the cases studied and dyskinesia in 14.2% of the cases. These findings suggest that drug-induced movement abnormalities are not limited to antipsychotics, but can result from SSRI antidepressants as well.

Neuroleptics

Of the classes of psychotropic medications listed in Table 5.1, antipsychotics (traditionally referred to as neuroleptics) contribute to the vast majority of untoward side effects that are likely to involve handwriting. Antipsychotics are prescribed to patients suffering from a variety of emotional, cognitive,

Table 5.1 Commonly Prescribed Psychotropic Medications and Their Mechanisms of Action, Indications, and Common Motor Side Effects (All Tradenames are Trademarked)

Generic Name	Trade Name	Mechanisms of Actions	Common Motor Side Effects
		Conventional Antipsychotics	
Haloperidol Chlorpromazine Fluphenazine Perphenazine Thioridazine Thiothixene Loxapine Trifluoperazine	Haldol Thorazine Prolixin Trilifon Mellaril Navane Loxitane Stelazine	Dopamine D2 receptor blockade; ranging from low potency (Mellaril) to high potency (Haldol)	Dystonia Akathisia Bradykinesia Tremor Dyskinesia
		Second-Generation Antipsychotics	
Clozapine Risperidone Olanzapine Quetiapine Aripiprazole Ziprasidone Paliperidone	Clozaril Risperdal Zyprexa Seroquel Abilify Geodon Invega	Less dopamine D2 blockade than conventional antipsychotics; also block or partially block serotonin receptors. Drugs vary in their DA/5HT receptor binding ratios	Akathisia Bradykinesia Tremor Dyskinesia
		Selective Serotonin Reuptake Inhibitor (SSRI) Antidepressants	
Citalopram Escitalopram Fluvoxamine Paroxetine Fluoxetine Sertraline	Celexa Lexapro Luvox Paxil Prozac Zoloft	Inhibits the reuptake of serotonin after being released in synapses; serotonin stays in the synaptic gap longer than it normally would and may repeatedly stimulate the receptors of the recipient cell	Akathisia Parkinsonism Dyskinesia
		Serotonin-Norepinepherine Reuptake Inhibitor (SNRI) Antidepressants	
Duloxetine Venlafaxine	Cymbalta Effexor	Same as SSRI but with added norepinepherine reuptake inhibition	Tremor Muscle weakness
		Other Antidepressants: Tricyclic Antidepressants; MAO Inhibitors	
Phenelzine Selegiline	Nardil Eldepryl	Blocks MAO, which breaks down excessive dopamine in the synaptic cleft	Actually reduced motor complications of levodopa therapy
Amitriptyline Busperone Bupropion	Elavil Buspar Wellbutrin	TCAs bind to serotonin and noradrenaline reuptake transporters to prevent the reuptake of these monoamines from the synaptic cleft, allowing their concentration to return to within the normal range	Akathisia Muscle twitches

(Continued)

Table 5.1 (Continued) Commonly Prescribed Psychotropic Medications and Their Mechanisms of Action, Indications, and Common Motor Side Effects (All Tradenames are Trademarked)

Generic Name	Trade Name	Mechanisms of Actions	Common Motor Side Effects
		Anxiolytics: Benzodiazepines	
Alprazolam Chlordiazepoxide Clonazepam Diazepam Lorazepam	Xanax Librium Klonopin Valium Ativan	Enhances the effect of the γ-aminobutyric acid (GABA) results in sedative effects	Excessive muscle relaxation; lack of coordination
		Mood Stabilizers	
Lithium	Lithium	Widely distributed in the central nervous system and interacts with a number of neurotransmitters and receptors, decreasing norepinephrine release and increasing serotonin synthesis	Tremor
Divalproex Lamotrigine Carbamazepine	Depakote Lamictal Tegretol	Anticonvulsant properties: targets the voltage-gated sodium channels and components of the GABA system	Tremor Parkinsonism

behavioral, sleep, and mood disorders. For this reason, we limited the discussion in remainder of this chapter to antipsychotic-induced side effects.

Neuroleptic medications have been the mainstay for treating major psychotic illness for over 50 years. The term *neuroleptic* comes from the Greek word *lepsis,* meaning a taking hold. This is an accurate description of the movement side effects that accompanied conventional neuroleptics. While neuroleptics improve the lives of schizophrenic patients, the occurrence of movement side effects, particularly parkinsonism, limited the therapeutic benefit of neuroleptics, so the treatment was often considered more problematic than the disease. Even after the emergence of a second generation of antipsychotics, drug-induced parkinsonism and dyskinesia continue to cause concern, particularly in vulnerable populations such as the elderly (Caligiuri, Rockwell, and Jeste 2000).

Antipsychotics are often prescribed to manage symptoms of schizophrenia (hallucinations, paranoia, agitation, and thought disorder), bipolar mania, psychotic depression, and agitation that often accompany dementia. Side effects from commonly prescribed antipsychotic medications include hyperlipidemia, weight gain, diabetes (collectively referred to as the metabolic syndrome), and motor problems such as parkinsonism and dyskinesia.

Motor side effects were common with first-generation (or conventional) antipsychotics such as haloperidol, fluphenazine, and chlorpromazine (see Table 5.1); however, with the advent of second-generation antipsychotics, the incidence of these troublesome side effects has decreased. Nonetheless, among elderly individuals, even on low dose, the newer antipsychotics can cause motor problems such as tremor, bradykinesia, and dyskinesia.

Antipsychotic-induced motor side effects may be classified in terms of their onset. Starting or switching to high doses of conventional antipsychotics often produces immediate reactions, usually within hours or days. The most common of these is dystonia, which is characterized by sustained muscle contractions. Common presentations of neuroleptic-induced dystonia consist of facial grimacing, tongue protrusion, throat tightness, torticollis, sustained open posture of the jaw, and abnormal or bizarre posturing of the trunk and limbs. Writer's cramp is an example of a dystonic reaction. Interestingly, acute dystonia is rare among older patients (Keepers and Casey 1991) but is 15 times more common in younger individuals (Raja 1998). Dystonic reactions have all but disappeared with the advent of second-generation antipsychotics; however, economic challenges within the national health care system are forcing some to reconsider the use of the inexpensive conventional agents as a cost-saving measure.

Acute reactions generally appear within days or weeks of starting an antipsychotic or increasing the dose of a current antipsychotic. Acute motor side effects include parkinsonism with all the classic motor signs (bradykinesia, rigidity, and tremor) and akathisia or restlessness. Drug-induced parkinsonism is nearly indistinguishable from idiopathic Parkinson's disease (PD). The two conditions stem from a similar mechanism involving reduced dopamine neurotransmission. In the drug-induced condition, motor signs appear following dopamine receptor blockade within the nigrostriatal pathway; in idiopathic PD, the condition develops following prolonged depletion of dopamine-producing neurons in the substantia nigra.

Perhaps the only observable feature that distinguishes drug-induced parkinsonism from PD is that in PD motor signs are usually asymmetric early in the course of the disease, whereas the signs are generally bilateral in drug-induced parkinsonism. This is because the offending agent in drug-induced parkinsonism does not favor one side of the brain over the other; it targets receptors bilaterally. In PD, the degenerative process is sporadic. Another important distinction between the two conditions is that drug-induced parkinsonism can be reversed by removing the offending agent.

Akathisia is one of the most common neuroleptic-induced extrapyramidal side effects. It emerges as part of treatment with either conventional or second-generation antipsychotics. Akathisia is observed in approximately 20%–40% of newly treated patients (Sachdev 1995). While its pathophysiology and epidemiology have attracted much attention over the past 40 years, our understanding of akathisia among older patients remains weak. Akathisia is characterized by a subjective feeling of restlessness or the urge to move and an objective motor component expressed as a semipurposeful movement most often involving the lower extremities. The movements have a driven quality to them; however, they are under voluntary control and can be suppressed for short periods of time except in extreme cases (Sachdev 1995). In mild akathisia, there is a subjective urge but the patient can control the urge to move and may suppress the unwanted motor activity. While use of second-generation antipsychotics has significantly reduced the incidence of drug-induced dystonia, parkinsonism, and tardive dyskinesia (Caligiuri et al. 2000), akathisia remains a problem (Kim and Byun 2010; Rummel-Kluge et al. 2011).

Longer term side effects are often persistent and irreversible. The most common persistent drug-induced motor side effect is tardive dyskinesia (TD), which is a syndrome characterized by choreoathetoid movements of the mouth, face, limbs, and trunk (Jeste et al. 1995; Yassa and Jeste 1992). Among patients treated with conventional antipsychotics, the lifetime prevalence of TD is reported to be 20%–25% (Yassa and Jeste 1992). This figure has decreased dramatically since the advent of second-generation antipsychotics; however, it has not been completely eradicated. As with other drug-induced motor side effects, with economic pressures encouraging wider use of conventional antipsychotics, we can expect an increase in the prevalence of TD.

The involuntary movements of TD generally appear in the hands and orofacial areas. This creates an embarrassing and uncomfortable situation for the patient and can have significant impact on daily life, including employment and socialization. Only in severe cases do the involuntary movements cause medical problems. Dyskinetic hand movements impair handwriting and other functions requiring fine control of hand and finger movements. As we will demonstrate in Chapter 11, handwriting movements in patients with TD (and even individuals treated with antipsychotics but without obvious dyskinesia) exhibit patterns of dysfluency with excessive changes in movement acceleration compared to normal healthy writers.

Neurobiology of Psychotropic-Induced Movement Disorders

Psychotropic medications produce a wide variety of movement disorders, primarily by altering normal neurotransmission within and through the basal ganglia. For the purpose of this chapter, we will limit the remaining section to a discussion of the pathophysiology of psychotropic-induced bradykinesia (slowness) and dyskinesia (excessive involuntary movements) due to their likely impact on handwriting. Acute dystonic reactions and akathisia are common following administration of a potent antipsychotic or increase in antipsychotic dose; however, other than writer's cramp, the persistent effects of these conditions on handwriting are unknown. In general, psychotropic-induced poverty of movements is thought to stem from disruption within the direct cortico–striato–pallidal pathway, whereas psychotropic-induced excessive movements are thought to stem from disruption within the indirect cortico–striato–pallidal pathway.

Localizing the neuronal circuits involved in parkinsonian bradykinesia has been the focus of animal and human research for over several decades. Primate models of basal ganglia disorders demonstrate that parkinsonian bradykinesia results from disruptions of normal inhibitory striatal projections to the internal segment of the globus pallidus (Albin, Young, and Penney 1989; Alexander, Crutcher, and DeLong 1990; Delong 1990). Excessive inhibition of outflow projections from the globus pallidus (internal; GPi) to the thalamus reduces the thalamocortical excitation, thereby reducing cortical excitation. This in turn would lead to a reduction (hypokinesia) or slowing (bradykinesia) of movement. Horak and Anderson (1984) and Mink and Thatch (1993) observed that monkeys with lesions causing increased activity within the GPi exhibit significant motor slowing. The therapeutic effects of surgical disruption of the output of the GPi in Parkinson's disease are also highly consistent with this model (Pfann et al. 1998; Alkhani and Lozano 2001; Lozano and Lang 2001; Dostrovsky, Hutchinson, and Lozano 2002).

Neuroanatomical and neurochemical bases for dyskinesia have been elucidated through studies of animal lesions (Alexander et al. 1990; DeLong 1990) and drug-induced dyskinesia in humans (Pahl et al. 1995; Brooks et al. 2000; Rascol et al. 1998; Henry et al. 2003). Dyskinesia can result from a loss of striatopallidal GABAergic inhibitory outflow to thalamic neurons causing increased thalamocortical excitation (Albin et al. 1989; Alexander et al. 1990). This model (see Chapter 1) assumes strong interactions between dopamine (at the level of the striatum), enkephalin (mediating GABAergic activity via dopamine D2 receptors), inhibitory GABA (within the globus pallidus), and excitatory glutamate (glutamate, at the level of the subthalamic nucleus).

Evidence supports a strong dopamine–GABA–glutamate interaction in the pathogenesis of dyskinesia. Specifically, a homeostatic dopamine–glutamate interaction has been shown to exist following repeated exposure to an indirect dopamine receptor agonist in laboratory animals (Canales et al. 2002). Repeated exposure to cocaine leads to prolonged decrease in dopamine release and reduced D2 receptor binding in the striatum and subsequent increased corticostriatal responsivity through an increase in glutamate release (Barretta, Sachs, and Graybiel 1999). Dopamine D2 receptors are associated with activation of GABA of the indirect circuit, the pathway implicated in dyskinetic movements (DeLong 1990).

Summary

In summary, while psychotropic medications offer therapeutic relief for a number of emotional, mood, and behavioral disorders, they are known to produce a wide range of undesirable motor side effects. Given the ubiquitous accessibility of psychotropic medications today, particularly in the aging population, it is important that the forensic document examiner gain an appreciation of the potential influence of these common medications on handwriting.

While any psychotropic agent has the potential to cause motor side effects, antipsychotics are more prone than other classes of drugs to affect handwriting movements. Antipsychotic-induced movement disorders may be grouped into acute or subacute conditions, such as dystonia, parkinsonism, or akathisia, as well as later occurring conditions such as tardive dyskinesia. This is important to know when evaluating handwriting samples that appear to reflect change in an individual known to be treated with an antipsychotic agent. Acute conditions such as dystonia appear within a few hours or days of starting an antipsychotic or increasing the dose of a previous antipsychotic. Dystonic reactions manifest as writer's cramp, limited range of movement (i.e., reduced stroke length), and fatigue.

Other acute reactions such as parkinsonism and akathisia (restlessness) generally appear within a few days or weeks of starting a new antipsychotic or increasing the dose. Parkinsonian manifestations would include micrographia (decreased stroke length), increased stroke duration, reduced stroke velocity, and possibly tremor. There are no known consequences of akathisia on handwriting. Tardive conditions such as dyskinesia have delayed onsets, sometimes taking months to appear. While acute side effects are thought not to persist as the patient usually develops a tolerance to the offending agent, tardive motor side effects can be persistent. Tardive antipsychotic-induced motor side effects usually are limited to dyskinesia, but can include later onset parkinsonism or akathisia. Table 5.2 summarizes the common drug-induced motor side effects, putative behavioral characteristics, and effects on handwriting.

Table 5.2 Common Drug-Induced Motor Side Effects, Their Behavioral Characteristics, and Effects on Handwriting

	General Behavioral Manifestations	Handwriting Movements
Dystonia	Sustained postures of arms, hands, tongue, eyes	Writer's cramp
Parkinsonism	Slowness, prolonged movements, tremor	Slowness, tremor
Dyskinesia	Random involuntary movements of face and hands	Jerky, lack of smoothness, increased number of acceleration peaks

Aging and Motor Control

6

Introduction

According to US census data, it is estimated that by the year 2030, 20% of US residents will be age 65 or older, reaching a population of over 88 million people by 2050. With advanced age there is decline is cognitive and sensorimotor function affecting fine motor control, balance, and gait. Routine daily activities become difficult. The decline in motor function stems from multiple factors, including alterations to both central and peripheral nervous systems governing neuromotor function (Seidler et al. 2010).

Age-related motor changes manifest in various forms, including tremor, which can reduce the ability to perform fine motor tasks; diminished postural reflexes, which can lead to loss of balance and injurious falls; and motor slowing, which can impact driving and other physical activities (Potvin et al. 1980; Kolb et al. 1998). The rate of occurrence and magnitude of these impairments varies substantially among individuals but typically develops gradually and may become sufficiently incapacitating to be considered pathological.

Converging evidence suggests that declines in striatal dopamine play a particularly important role in age-related motor declines (e.g., Bannon et al. 1992; Carlsson 1981; Haycock et al. 2003; McGeer, McGeer, and Suzuki 1977; Mozley et al. 1999). In addition to the findings from correlational studies, research also shows that exposure to a range of pharmacological agents that destroy, block, or diminish striatal dopamine neurotransmission produce the motor changes that are often present in advanced age (Betarbet, Sherer, and Greenamyre 2002; Di Monte, Lavasani, and Manning-Bog 2002; Langston et al. 1983).

In this chapter we first summarize the biochemical evidence supporting an important role of dopamine in the genesis of age-related motor decline. Following this review, we provide an overview of the specific motor deficits commonly observed in an aging population. Lastly, we focus on the relevance of advanced age to specific motor behaviors that could impact handwriting. The overall goal of this chapter is to enable the reader to make direct inference from observations of impaired handwriting in an elderly writer to natural age-related alterations in neural substrata governing fine motor control.

Neurotransmitter Mechanisms of Motor Aging

Aging and Nigrostriatal Neuronal Cell Loss

The most extensive postmortem studies of human striatal dopamine were published by Carlsson (1981) and Carlsson and Winblad (1976). Their most significant finding was that the rate of dopamine loss accelerated after age 60. Others have replicated these findings. For example, Bugiani et al. (1978) calculated a 70% loss of nigrostriatal neurons after the age of 55, and McGeer and McGeer (1978) found a 66% loss in nigral dopamine among individuals aged 50–90 compared to those aged 18–30. The nigrostriatal system has been implicated in age-related motor decline as well (Umegaki, Roth, and Ingram 2008). McGeer (1978) conducted a study of neuron count in postmortem brains of individuals aged 18–30 compared with those of individuals aged 50–90. She reported a reduction in the number of neurons in the substantia nigra from 380,000 for the younger group to 250,000 for the older group, a loss of 66%.

Similar reductions in age-related neuron count were reported by Brody (1955) and Bugiani et al. (1978). Figure 6.1 is adapted from the data published by Brody and shows the age-related decline in neuron count for the striatum and precentral motor strip. One can readily see that the neuronal loss in the striatum is not linear; rather, the greatest loss occurs during midlife with a gradual decline throughout senescence. Also relying on postmortem brain tissue, Bugiani et al. reported a 70% reduction in neuron count in the putamen over the age range from 19 to 65 years.

Normal motor function depends on a balance between dopamine and acetylcholine in the striatum (see Chapter 1 for an overview of neurotransmitters and striatal function). Advanced age unequivocally leads to a decrease in the concentration and binding of dopamine D2 receptors in the striatum (Carlsson and Winblad 1976; Morgan and Finch 1988; Umegaki et al. 2008). Recall from Chapter 4 that the principal neuropathological mechanism underlying Parkinson's disease is the reduction of nigrostriatal dopamine transmission. It is no coincidence that the motor characteristics observed in Parkinson's disease are more severe forms of motor impairment (that appear earlier in life) found in healthy aging.

Dopamine neurotransmission is altered at both molecular and cellular levels in advanced age. At the molecular level, the biosynthesis of D2 receptors is reduced, thus impacting the number of dopamine binding

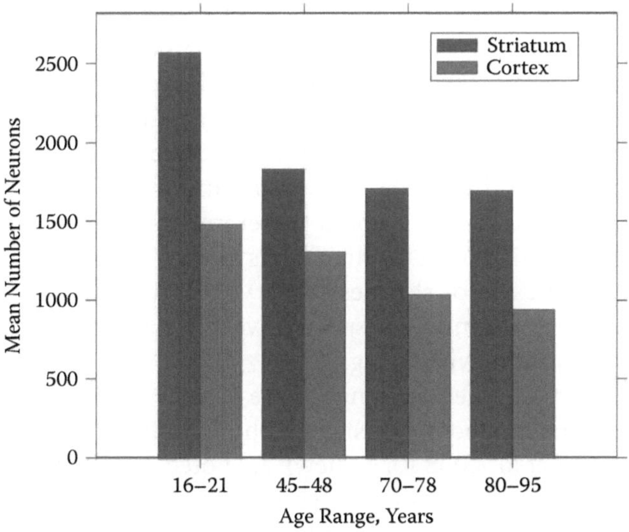

Figure 6.1 Neuron count in two brain areas estimated from postmortem brains of individuals in four age groups. (Adapted from Brody, H. 1955. *Journal of Comparative Neurology* 102:511.)

sites (Henry and Roth 1984). At the cellular level, approximately 20% of striatal neurons disappear with advanced age (Han et al. 1989). As the number of striatal neurons diminishes over time, there is a natural (autogenic) attrition of receptor sites (i.e., "use it or lose it"). This leads to an effective loss of approximately 30%–50% in dopamine D2 receptors within the striatum. Given that these receptors are the primary targets for nigrostriatal neurons, it is not surprising that the motor consequences of these striatal changes bear a striking similarity to Parkinson's disease—suggesting to some that the age-related decrease in striatal dopamine function is part of the preclinical continuum of Parkinson's disease (Romero and Stelmach 2001).

In a recent study of age-dependent changes in dopaminergic neuron firing patterns, Ishida et al. (2009) reported that, in animals, the normal distribution of firing patterns is altered in aging. Dopamine neurons within the substantia nigra pars compacta exhibit three modes of firing: pacemaker, random, and burst. These three modes vary depending on afferent modulation from other basal ganglia nuclei. In the presence of GABAergic input to the striatum, the dopamine neuron firing pattern changes from pacemaker mode to burst mode (Lee and Tepper 2009). Ishida noted that, in the aging rat, the firing pattern of dopamine neurons changed from pacemaker to random, and then to burst mode. Given the Lee and Tepper (2009) findings, this would suggest that advanced age may lead to an unregulated increase in GABAergic input to the striatum.

Recall from Chapter 1 that excitatory dopamine D1 receptors are found on nerve terminals that are part of the direct striatopallidal pathway, whereas the inhibitory dopamine D2 receptors are found on nerve terminals that project within the indirect pathway. Both function to decrease thalamocortical inhibition and thus facilitate movement. An increase in GABAergic inhibition (as suggested by the Ihsida et al. 2009 finding) of receptors terminating on excitatory D1 receptors would therefore have a net effect of reducing movement. While this causal mechanism is likely oversimplified, it may very well account for the motor slowness observed in the aging animal. Further research is needed to link this speculative mechanism to the behavioral observations associated with aging.

Aging imparts another challenge to the healthy striatum. Several animal studies demonstrate a decline in striatal cholinergic activity with increased age (see Umegaki et al. 2008 for review). As noted earlier in Chapter 1, normal motor function depends on a critical balance between the inhibitory influence of dopamine and excitatory influence of acetylcholine within the basal ganglia. This cascade of events impairs the reciprocal inhibitory control between dopamine and acetylcholine, leading to impaired motor function.

Advanced age impacts dopamine synthesis, transport, and binding within the striatum. In the following paragraphs we summarize the literature on how these three mechanisms are altered in advanced age.

Aging and Dopamine Transporter Mechanisms

Dopamine transporter (DAT) is an integral membrane protein that removes dopamine from the synaptic cleft and deposits it into surrounding cells. DAT enables the transmission of dopamine from one nerve to another within the dopamine pathways. Not unexpectedly, the natural aging process imparts significant reductions in DAT. Our current understanding of age-related changes in DAT in humans derives from studies of postmortem brain tissue (Bannon et al. 1992; Haycock et al. 2003) and functional neuroimaging techniques (Mozley et al. 1999; van Dyck et al. 2002; Volkow et al. 1996). Bannon and colleagues (1992) found that DAT mRNA levels in substantia nigra were relatively constant through the age of 57, after which levels declined by 95%. It is not clear from their report whether the oldest subjects in the study had histories of motor abnormalities.

Haycock et al. (2003) measured levels of DAT from postmortem human striatum and found a significant 13% decline in caudate during aging. Positron emission tomography (PET) and single-photon emission computed tomography (SPECT) techniques have also been used to elucidate the effect of aging on DAT. Using PET, Volkow et al. (1996) found that DAT availability decreased significantly after the age of 40 in the caudate and putamen, declining about 6.6% per decade of life in normal healthy individuals. Mozley and colleagues (1999), using SPECT, found a nonlinear decrease in DAT uptake sites across the age range in both the caudate and putamen, and van Dyck et al. (2002) reported a 45%–48% decline in DAT over the age range from 18 to 88 years in the putamen and caudate, respectively. In a more recent study, van Dyck et al. (2008) showed that the decreased level of DAT was associated with increased simple reaction time in older adults. Interestingly, they reported a 6.6% decline in DAT per decade, a figure that agrees with the results of Volkow et al. (1996) using PET.

Dopamine Neurotransmission

Morgan and Finch (1988) concluded that while striatal dopamine levels do change with age, these changes might be secondary to loss of nigral cells (P. McGeer et al. 1977) or loss of tyrosine hydroxylase activity, an important enzyme for dopamine neurotransmission (McGeer and McGeer 1976). Haycock et al. (2003) addressed the discrepancy between the dramatic reduction of nigral dopamine with age and the relative stability of presynaptic dopamine markers, suggesting that the dopaminergic system appears capable of compensating for neuronal loss during aging in healthy individuals.

Figure 6.2, adapted from McGeer and McGeer (1976), shows the percentage decrease in various striatal enzymes responsible for dopamine neurotransmission. Using postmortem brain tissue, the McGeers estimated

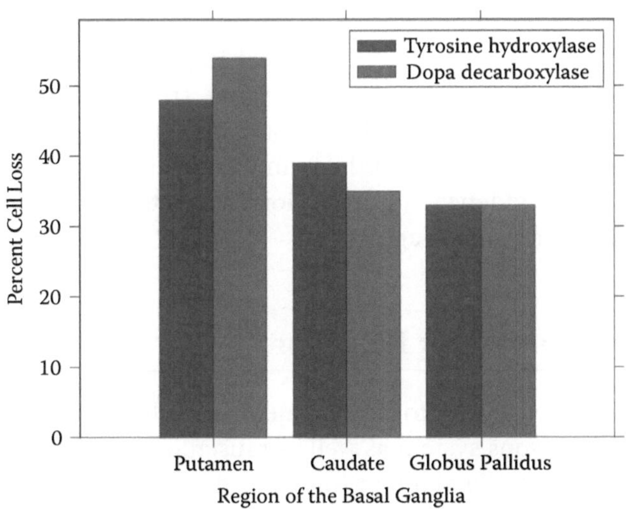

Figure 6.2 Percent reduction in dopamine enzymes from age 25 to 50. (Adapted from McGeer, P. L., and McGeer, E. G. 1976. *Journal of Neurotransmission* 26:65–76.)

dopamine cell loss based on the percentage loss of these enzymes in the putamen, caudate, and globus pallidus from age 25 to 50.

Aging and Dopamine Receptor Changes

Studies examining the effects of age on dopamine receptor properties have employed both neuroimaging techniques and postmortem tissue analyses. Wong et al. (1988) used PET to examine caudate D2 dopamine receptors in subjects aged 19 to 73. They found a significant negative correlation between age and D2 receptor binding in the caudate, whereas DeKeyser, Ebiner, and Vanquelin (1990) failed to find a relationship between age and D1or D2 receptor concentrations in the putamen based on postmortem tissue analyses. However, they did observe a significant negative relationship between age and concentration of dopamine uptake sites residing on the dopamine nerve endings in the putamen (r = –0.89).

In two more recent PET studies, dopamine D1 and D2 receptor concentrations in the caudate and putamen were found to decline with age. Wang et al. (1998) reported a 6.9% decrease in D1 receptor binding in the caudate and a 7.4% decrease in D1 receptor binding in the putamen per decade over an age range from 22 to 74 years. Volkow et al. (1998) reported significant effects of advanced age on dopamine D2 receptor availability in the caudate and putamen. Moreover, they found that performance on a finger-tapping task was correlated with D2 receptor availability in the caudate (r = 0.66) and putamen (r = 0.66), such that faster tapping rates were associated with greater D2 receptor concentrations.

In summary, the normal aging process imparts significant changes in dopamine neuronal markers. Neuroimaging studies show striatal DAT decreases approximately 7% per decade or nearly 50% over the average adult life span. The rate of loss of striatal dopamine appears to accelerate after age 55. Lastly, there is evidence of a reduction in striatal dopamine D1 and D2 receptor sites with advanced age.

Aging Effects on Motor Behavior

The cause of age-related motor impairment is likely multifactorial. Central and peripheral nervous system as well as musculoskeletal factors contribute independently to functional motor impairments that accompany senescence. Central nervous system factors are of particular relevance to complex movements such as handwriting. The natural aging process is accompanied by alterations in brain anatomy known to play important roles in the planning and execution of complex motor behavior.

For example, investigators have reported decreases in cerebral volume, ventricular dilatation (Davis, Mirra, and Alazraki 2005; Schretlen et al. 2000), and declines in regional blood flow in the prefrontal and frontotemporal cortices, thalamus, putamen, and caudate (Melamed et al. 1980) in older, otherwise healthy individuals. As inferred from the previous discussion on striatal cell loss, one of the most predominant findings in normal aging is the reduction in the width of the substantia nigra (Pujol, Junque, and Vendrell 1992). Alterations in the substantia nigra are reported in normal aging and result in diminished striatal dopaminergic function. In normal aging, declines in dopamine neurotransmission are correlated with poor performance on a variety of motor tasks (Pujol et al. 1992).

Functional declines in motor behavior due to aging generally stem from impaired coordination (Seidler, Alberts, and Stelmach 2002), increased variability in movement trajectory or muscle force (Contreras-Vidal, Teulings, and Stelmach 1998; Darling, Cooke, and Brown 1989), or slowing of movement (Diggles-Buckles 1993). Interestingly, impairment in each of these domains can have a profound impact on handwriting movements. In the following paragraphs, we briefly summarize the literature on relevant findings on the effects of age on motor behavior. Motor impairments observed in the elderly may be grouped into clinical signs often associated with or predispose an individual to a particular neurological diagnosis (e.g., Parkinson's disease) or nonspecific motor disturbances that reflect more basic motor pathophysiology. The clinical motor signs would consist of tremor, hypertonic, and diminished postural reflexes. These problems are seen in both neurological disease and normal aging. The nonspecific motor signs would consist of kinematic impairments such as increased reaction time, decreased speed, increased movement time, increased variability, and reduced grip strength.

Clinical Motor Manifestations of Aging

Increased Muscle Tone

Muscle stiffness is a frequent complaint among older individuals. Bennet et al. (1996) reported that approximately 45% of community-dwelling individuals over the age of 85 exhibited signs of upper extremity rigidity. However, Kaye et al. (1994) were unable to detect significant differences between healthy individuals with a mean age of 79 and those with a mean age of 89 on their measure of upper extremity muscle tone, suggesting that age alone may not account for marked increases in hypertonia. It is possible those changes in muscle tone consistent with nigrostriatal changes in dopamine are nonlinear throughout the age range or that they begin to appear during middle age and stabilize during the latter years of life in the absence of pathology.

Tremor

Tremor may be the most ubiquitous motor sign associated with advanced age. According to Louis et al. (1998), "[T]remor can be clinically detected in almost all individuals at every age, even when subjects are unaware of having a tremor" (p. 225). The earliest report we could find quantifying the effects of aging on physiological tremor was a study by Marshall (1961). Using accelerometry to record tremor frequency from the hand, Marshall reported a significant decline in tremor frequency from a mean of approximately 9 Hz to a mean of 6.5 Hz over the ages of 30 to 90. Based on studies showing that tremor frequencies between 8 and 10 Hz reflect normal physiological tremor, whereas those between 4 and 7 Hz are typically associated with pathology (Elble and Koller 1990), the Marshall findings indicate that during the aging process, tremor evolves from a normal physiological state to one resembling parkinsonism.

More recent studies support this conclusion. For example, Wade, Gresty, and Findley (1982) reported that the modal tremor frequency among individuals up to 70 years of age was relatively constant at 7 Hz and declined thereafter to about 6 Hz. Elble (2003) used electromyography (EMG) to examine motor-unit entrainment in young and elderly subjects. While the results were inconclusive with regard to prevalence of tremor peaks across the age range, when present, the EMG peak frequency was 9–12 Hz in younger subjects and 5–7 Hz in older subjects.

Large-scale prevalence studies indicate that tremor severity increases with age (Louis et al. 1998). Moreover, based on studies of the spectral properties of tremor, it may be concluded that advanced age is associated with lower tremor frequency. As such, the presence of a low-frequency (4–7 Hz) resting tremor, particularly in an older individual, may be a harbinger of neurological disease.

Postural Reflexes

Advanced age disrupts gait and balance (Elble et al. 1991; Maki, Holliday, and Fernie 1990; Wolfson et al. 1992). In addition to central postural reflexes, postural control is accomplished through the integration of somatosensory and visual feedback. Older individuals have slower and less reliable postural reflexes than younger individuals (Stelmach et al. 1989). King, Judge, and Wolfson (1994) reported that older individuals standing on a sway platform were less able to compensate for sudden perturbations. Reflexive responses declined 33% from the third to the eighth decade of life. In a comprehensive study of neurological function in normal individuals aged 64–100, Kaye et al. (1994) found abnormally diminished reflex responses in as many as 29% of the subjects younger and 56% of the subjects older than 80 years of age. The most prevalent reflex abnormality involved the gastrocnemius-soleus reflex, the primary reflex involved in postural stability.

Diminished postural reflexes are the primary contributors to functional gait and balance disturbances. Gill et al. (2001) and Du Pasquier et al. (2003) reported age-related reductions in balance when subjects were examined using standard assessments. Although few studies have examined effects of age on postural balance longitudinally, Era et al. (2002) reported that postural balance, measured by normal standing with eyes closed, deteriorated significantly over the 5 years between age 75 and 80.

Nonspecific Age-Related Motor Impairments

Reaction Time

Since Galton's 1899 publication on reaction times (RTs) across the age spectrum, RT has been considered one of the more reliable indices of motor aging. Modern studies show a gradual linear increase in RT over age, beginning at approximately 20 years of age (Fozard et al. 1994; Wilkinson and Allison 1989). Potvin et al. (1980) studied 61 normal men ranging in age from 20 to 80, calculated the percentage decline in function on several motor measures, and found a 28% increase in simple reaction time.

Movement Duration

Unlike RT, which places demands on the attention system, movement time (MT) is associated with minimal cognitive load. As such it may be viewed as a measure of motor execution. Investigations of age-related changes in movement time show a 23% decrease in hand tapping speed from age 20 to 80 (Potvin et al. 1980). Kaye et al. (1994) studied motor functions in two groups of older community-dwelling individuals: young old (with a mean age of 70) and oldest old (with a mean age of 89). They reported significant reduction in the number of finger taps per second in the oldest old subjects. The investigators did not include younger subjects for comparison. Nonetheless, increased movement time has been a consistent finding of motor aging.

Diggles-Buckles (1993) reported that older adults increased their movement duration on a variety of tasks by as much as 30%. It has been argued that older adults increase movement time in order to maintain movement accuracy (Seidler-Dobrin, He, and Stelmach 1998). Alternatively, increased movement time may have a cognitive basis (e.g., difficulty managing attentional demands); however, research has shown that increasing attentional demands does not differently impact movement time in older compared to younger adults (Salthouse 1993; Salthouse and Somberg 1982).

Force Variability

Increased variability in motor performance has been a consistent finding in studies of aging and motor control. Of particular relevance to hand motor control is variability in force output, as many functional tasks involving the hand require greater precision in force than motion control. It has been hypothesized that the age-related increase in force variability is a result of variability in the discharge properties of single motor units (Galganski, Fuglevand, and Enoka 1993; Kamen and Roy 2000; Laidlaw, Bilodeau, and Enoka 2000). Several studies report increased force variability among older adults while they are performing isometric finger force tasks (Vaillancourt, Larsson, and Newell 2003; Tracy et al. 2005; Sosnoff and Newell 2006; Christou 2011; see Diermayr, McIsaac, and Gordon 2011 for review), particularly for low levels of force.

Vaillancourt et al. (2003) compared three groups of subjects with mean ages of 22, 67, and 82, respectively. The investigators found a somewhat linear increase in force variability with age on the static force task; however, on a more demanding sine-wave tracking task, older adults exhibited dramatically increased variability compared with younger subjects. Tracy et al. (2005) reported that the degree of force variability during a static isometric finger force task was associated with the variability on motor unit discharge, suggesting that a pure motor mechanism (Eisen, Entezari-Taher, and Stewart 1996; Enoka et al. 2003)—rather than cognitive (inattention) or sensorimotor (visual feedback) mechanisms—may be responsible for the age-related increase in fine motor fluctuation.

However, Sosnoff and Newell (2006) proposed that age-related increases in force variability might be due to decrements in strength rather than central processes. They reported that while older adults did indeed exhibit greater variability during maintenance of submaximal static force tasks compared to younger adults, once they controlled for strength (i.e., covarying for maximum voluntary contraction), the age-by-variability relationship disappeared. Instead, they found a strong inverse relationship between strength and force variability that was independent of subject age. They concluded that the observed age-related changes in force variability more fundamentally reflect weakness-related variability than an independent age effect.

Work by Cole and colleagues (Cole 1991, 2006; Cole and Beck 1994; Cole, Rotella, and Harper 1999) conveys a perspective on compensatory adjustments to grip force that older adults may use to offset declining tactile sensibility that often accompanies advanced age. In a series of innovative experiments involving pinch force, these investigators found that in the presence of reduced tactile information, older adults increased their grip force to prevent slippage of an object beyond what would be predicted by the scalar decrease in sensory information. Unlike the aforementioned studies reporting increased variability in maintenance of static finger force (e.g., Vaillencourt et al. 2003), Cole and Beck (1994) failed to find an age-dependent increase in pinch force variability. Instead, they observed that older adults (aged 68–85) produced higher levels of force than younger subjects to maintain precision grip. Despite the relative stability in maintaining precision force, these findings suggest that older adults employ a strategy that involves increasing their force to levels sufficient to offset age-related fluctuations in steady-state grip force.

The Vaillencourt et al. (2003) and Cole and Beck (1994) findings of a disproportionate increase in force variability and compensatory force level, respectively, have implications for evaluating handwriting among older adults. At least two features of handwriting may be influenced by an age-related increase in force variability. First, assuming that pen movements during normal handwriting require the writer to maintain relatively constant pen grip force throughout the extended series of dynamic, sinusoidal pen strokes, variability in grip force could adversely affect the smoothness and speed of pen stoke trajectories. Second, as we will discuss in greater detail in Part 2 of the book, pen pressure is an important feature that discriminates authentic from forged or disguised signatures. Document examiners employ procedures to infer pen pressure from careful examination of the depth and width of the static indentations present on the writing surface. Age-related variability in production and maintenance of low levels of force typically used during handwriting will manifest as variability in the depth and width of these static indentations, further challenging the examiner's ability to judge writer authenticity.

Grip Strength

Grip strength is considered an important predictor of longevity, general health, and quality of life throughout adulthood (Taekema et al. 2010; Rantanen et al. 2011). In the present context, decline in grip strength with age can inform the scientific and forensic communities tasked with understanding sources of variability in fine motor control and handwriting, respectively. Nahhas et al. (2010) reported that, on average, men and women attain maximum handgrip strength at age 36. Women, however, begin to show decline in grip strength at an earlier age (50 years on average) than males (56 years on average). Handgrip strength declines by approximately 1% per year after midlife (Rantanen et al. 1998). Individual variability in the rate of decline is likely determined by genetic factors (Silventoinen et al. 2008). A particularly interesting study by Lindberg et al. (2009) showed that older adults reach target grip force at a slower rate than younger adults, suggesting an age-related disturbance in the time necessary to recruit a sufficient number of motor neurons necessary to match the target force levels.

Motor Coordination

Thaler (2002) proposed that healthy aging is associated with a reduction in the available system states and responses. Translating this to movements having multiple degrees of freedom, this would predict that advanced age negatively impacts one's ability to organize the inherent kinematic variability optimally, as suggested by the principle of motor equivalence (see Chapter 3). Latash, Shim, and Zatsiorsky (2004) and Latash et al. (2006) suggest that the reduction in motor abundance observed in aging may be an adaptive response to loss of sensorimotor fidelity and increase in neuromuscular noise. As noted earlier, the elderly generally produce movements that are delayed at the onset, are longer in duration, and involve excessive muscle contractile force (particularly for the antagonist muscles) to ensure accuracy. These adaptive strategies have the effect of reducing the number of available solutions to achieving a goal.

The infinite combinations of digit forces and movement trajectories that are normally available to execute a complex multijoint movement such as handwriting are significantly reduced in aging (Seidler et al. 2002; Shinohara et al. 2003, 2004; Newell, Vaillancourt, and Sosnoff 2006; Lipsitz 2004; Verrel, Lovden, and Lindenberger 2010). Seidler et al. (2002) reported that age-related impairments in smoothness and accuracy were more pronounced for multijoint than single-joint movements. In a study of manual pointing, Verrel et al. (2010) found no significant differences between younger (mean age 25.5) and older (mean age 73.4) adults on measures of movement duration, peak velocity, or pointing accuracy. However, they observed that older adults tended to attain peak velocity earlier than younger adults.

On tasks requiring a rapid single movement trajectory to reach a specific endpoint, the movement profile typically has two phases: a rapid ballistic phase associated with movement initiation and a slower "homing in" phase associated with attaining accuracy. Location of the peak velocity separates these two phases. In older adults, peak velocities occurring earlier in the trajectory suggest a prolonged "homing in" phase and greater reliance on visual feedback (Seidler-Dobrin, He, and Stelmach 1998).

Summary

Normal aging is accompanied by a significant reduction in dopamine neurotransmission, particularly in the striatum leading to alterations in motor function that, in advanced aging, resemble early parkinsonism. Age-related declines in motor function manifest as initiation delay, reduced speed and increased movement duration, increased force variability, and loss of coordination of multijoint synergies. Advanced age compromises one's ability to organize the inherent kinematic variability optimally and execute a desired movement sequence. Despite this bleak outlook, the human motor system is a highly redundant system endowed with multiple degrees of freedom offering options for the aging adult to adapt to and partially compensate for these sensorimotor and neuromuscular deficits.

With regard to handwriting motor control, certain age-related impairments will have more deleterious effects than others. Specifically, among the clinically relevant motor signs, tremor will clearly impact handwriting movements and reveal stroke dysfluencies and oscillations. A writer's effort to inhibit tremor by increasing muscle stiffness will result in restricted movements and reduced stroke amplitudes. Among the nonspecific motor effects of aging, increased movement duration, increased variability, and reduced grip strength will alter both qualitative and quantitative aspects of handwriting and can be readily observed from the static hard-copy documents. The problem of variability is of particular significance to the document examiner. Fluctuations in force steadiness and inconsistent deployment of adaptive strategies can introduce variability in many features of the handwriting movement, including amplitude, slant, smoothness, and pen pressure. More importantly, these fluctuations can occur within a single document and over time between documents.

The goal of this chapter was to provide background on the fundamental neurochemical changes that accompany aging and to describe the functional consequences of aging on motor control in general and briefly broach the topic of handwriting. In Chapter 13, we summarize the available literature on the effects of healthy aging on handwriting.

Kinematics of Signature Authentication

II

A Kinematic Approach to Signature Authentication

7

Introduction

The aim of Section II is to describe a quantitative approach to the dynamic analysis of handwriting and signatures. While the vast majority of research on signature authentication has focused on static traces, modern technology has enabled researchers to quantify the kinematic features of signatures at the level of an individual pen stroke. Historically, visually detectable features in handwritten signatures formed the basis of evidence supporting whether a questioned signature is genuine, disguised, or forged (Osborn 1929; Hilton 1961; Michel 1978; Herkt 1986; Mohammed 1993; Huber and Headrick 1999; Wendt 2000; Durina 2005; Mohammed et al. 2011). Today, research into static features associated with different signing behaviors can be supplemented by dynamic studies where kinematic data are collected from signatures recorded on digitizing tablets. This technique has been used to report on the effects of disguise and simulation behaviors in terms of pen pressure, stroke formation, and movement duration (e.g., van Gemmert et al. 1996).

This chapter provides an overview of the features and parameters that can be extracted from signatures and handwriting samples using this approach. In general, the analysis utilized computer software to digitize and extract multiple kinematic variables from each pen stroke. Examples are provided from previously published and ongoing research from our laboratories on the kinematic analyses of genuine, disguised, and forged signatures to demonstrate the application of this approach. In Chapters 8 and 9 we describe results from research designed to test whether a given signature is the product of highly programmed motor behavior (i.e., authentic) or a forgery (i.e., an attempt to "overwrite" an internal handwriting program) to be tested in practice.

The early forensic document examiner (FDE) pioneers, such as Albert S. Osborn, established their roots in the teaching of penmanship. These FDEs were skilled penmen themselves and they worked in a time when handwriting was taught as a necessary skill for business. They therefore were experts on the handwriting systems of the day and of the past. They could tell when deviations were made from the various copybook systems. They referred to the copybook styles as class characteristics and the deviations as individual characteristics. Their system of handwriting identification was based on distinguishing individual characteristics and determining whether they were written by one writer or two, or whether there had been an attempt to simulate a person's handwriting characteristics.

The vast majority of research by FDEs regarding signatures focused on static traces. The classic FDE texts described the features of genuine signatures as flying stops and starts, variation in pen pressure or pen load, speed, and good line quality (Osborn 1929; Harrison 1958; Hilton 1961; Conway 1959; Huber and Headrick 1999; Morris 2000; Seaman-Kelly and Lindblom 2006). However, FDEs must infer kinematic information about duration, speed, pen pressure, and tremor from static traces (Guest, Fairhurst, and Linnell 2009). A quickly written signature with variation in pen pressure and little dysfluency indicates authenticity if the specimen signatures display the same qualities. On the other hand, a slowly written, shaky signature with little variation in pen pressure is evidential of a simulation or forgery.

However, as the teaching of handwriting as a skill has become less of a priority in schools and the intermovement of populations increased, the use of handwriting systems as the basis for handwriting identification has become less useful. A more contemporary FDE view is that "the possibility of identifying the particular system behind the writing of any individual of North American origin today is extremely remote" (Huber and Headrick 1999). This position is further supported by research on the variety of handwriting systems being taught in Canada today (Holmes 2010).

Found and Rogers (1999) state that "under normal conditions, given a sufficient amount of writings, no two skilled writers are likely to produce handwritten images that are exactly the same in terms of the combination of construction, line quality, formation variation and text structure features." Harrison (1958) stated that "there is no doubt that on the basis of letter design alone, the number of distinguishable handwritings is virtually unlimited for all practical purposes." Unlike Harrison, Found and Rogers base their observations on motor control rather than class and individual characteristics derived from copybook systems.

An alternate approach to handwriting identification was proposed by Found and Rogers in 1998. Their feature detection and complexity theory is based on neurobiological principles. They considered that the complexity of a signature was a product of a combination of the formation, concatenation, and intersection of the strokes and number of turning points that comprised the signature. They hypothesized that the more complex the signature was, the harder it would be to simulate and the less chance there would be of a chance resemblance (Found and Rogers 1996, 1998).

There have been some recent attempts to test the complexity approach. A team of Dutch researchers tested several FDEs and, using their data set, derived equations to calculate the complexity of a signature (Alewijnse et al. 2009). This work is ongoing and may possibly lead to the development of a complexity scale, which would increase the objectivity of FDEs.

The Kinematic Approach

Early attempts to transform complexity theory to practical application for quantifying signature characteristics involved manually counting the number of intersections and retraces associated with a given signature (Found et al. 1998). Figure 7.1 shows two examples illustrating the application of this method used manually to count the number of intersections and retraces associated with each signature.

It is important to reiterate at this point that FDEs examine static traces. That is, the signature when examined is in the form of an ink trace on the substrate, normally paper. Because of this, significant dynamic information (such as stroke duration and velocity) is lost to the examiner. The kinematic

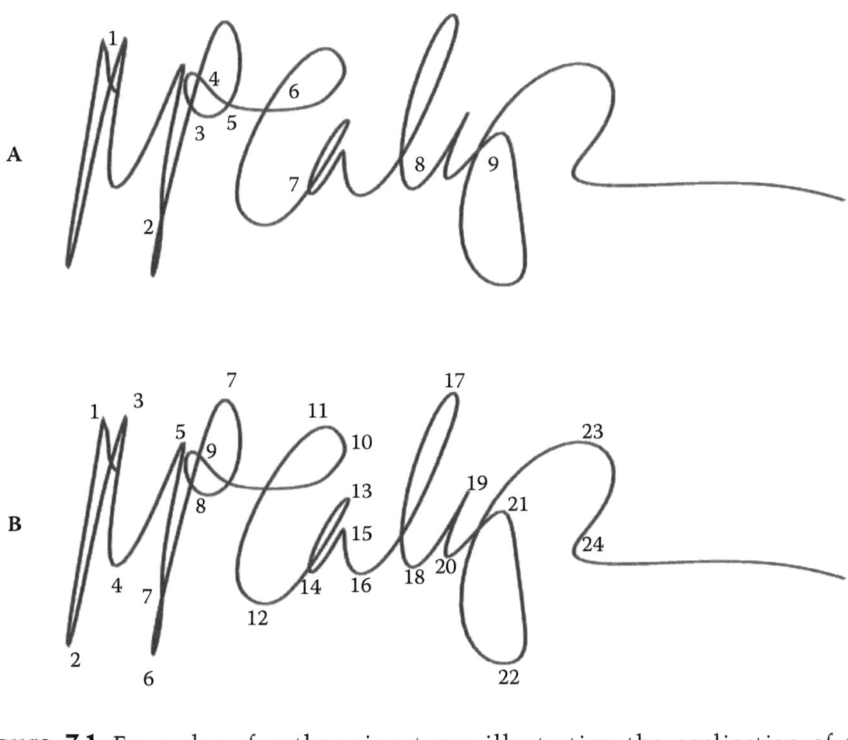

Figure 7.1 Examples of author signatures illustrating the application of the method used manually to count the number of intersections (a) and retraces (B) associated with each signature.

approach to signature and handwriting examination involves the development of databases of signatures and handwriting that are collected dynamically. Research involving dynamically written signatures and handwriting is usually undertaken under different conditions (genuine, disguised, and simulated) by healthy writers and writers who are compromised by conditions that may affect their handwriting.

Modern kinematic approaches utilizing digitizing tablets such as those marketed by Wacom[1] combined with the use of software such as MovAlyzeR[2] are very powerful tools in collecting dynamic data. Figure 7.2 illustrates graphs produced with the use of MovAlyzeR showing extraction of velocity (B) and pen pressure (C) over time.

The resulting databases can then be statistically analyzed to determine interactions between writing styles and writing conditions. This information will provide FDEs with empirical data that will assist them in their evaluations of kinematic information from static signatures.

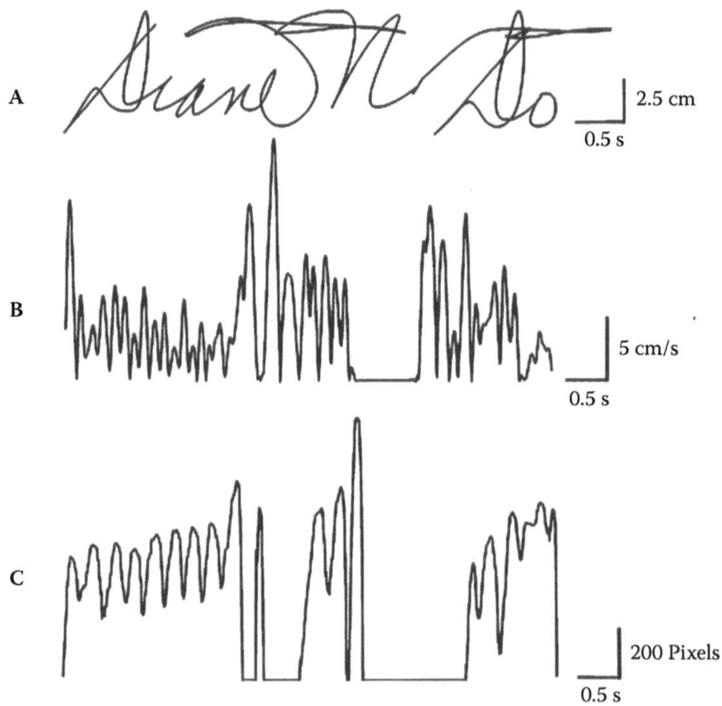

Figure 7.2 Sample signature and processed waveforms produced using MovAlyzeR software. Shown are the (A) unprocessed signature, (B) absolute velocity waveform over time, and (C) pen pressure over time with amplitude and time calibration bars. Note that pen lifts are recorded in the unprocessed signature and velocity trace, but register as zero pressure in the pressure trace.

Kinematic Methods

The kinematic approach begins with a digitally recorded handwriting sample. Several digitizing tablets are commercially available for this purpose. Forensic applications generally utilize a special inking pen for this purpose, while many scientific applications (see Chapters 10–12) utilize noninking pens to minimize error correction and control visual feedback. It is important to consider the sensitivity (resolution) and sample rate of the digitizer when a tablet is selected. Software is needed to acquire and process the handwriting samples. The forensic and scientific research conducted in our laboratories is based on signature and handwriting samples recorded and processed using a Wacom digitizing tablet (see Figure 7.4 later in the chapter) and MovAlyzeR software.

Once the samples are recorded and stored on a computer, software can automatically segment pen movements into successive up and down strokes using interpolated vertical-velocity zero crossings. The basic unit of movement in which we are interested is therefore the stroke. Our research focuses on the vertical movement component only because this is the main movement component in Western cursive handwriting and handprint. Table 7.1 shows a list of the dynamic variables commonly extracted from each segmented stroke (although more are available).

These features are calculated for the primary and secondary submovements (Meyer et al. 1988). The primary submovement begins where the stroke begins and ends where the vertical velocity changes from decelerating to accelerating for the first time after the velocity peak. The primary submovement is comparable to the initial, ballistic phase of the up or down stroke. Thus, acceleration peaks in the primary submovement occur before the velocity peak, while the total number of acceleration peaks can occur before or after the velocity peak. Secondary submovements are associated with the final adjustments (or "honing in") and corrective movements.

Table 7.1 Dynamic Variables Commonly Extracted from Each Segmented Stroke during the Analysis and Summarization Process

Stroke duration, in ms
Stroke length, in cm
Peak stroke velocity, in cm/s
Average stroke velocity, in cm/s
Time to peak velocity, in ms
Stroke peak acceleration, in cm/s/s
Number of acceleration peaks
Average normalized jerk
Pen pressure

Two of the variables shown in Table 7.1 are used to quantify smoothness or fluency of pen movements. These are the number of acceleration peaks (or inversions) and average normalized jerk (ANJ). Normalized jerk is unitless as it is normalized for stroke duration and length. Average normalized jerk is calculated using the following formula: $\sqrt{}(0.5 \times \Sigma(\text{jerk}(t)^2) \times \text{duration}^5/\text{length}^2$ (Teulings et al. 1997). Higher ANJ scores and increased number of acceleration peaks per segment are indicative of dysfluent writing movements or dyskinesia.

Additionally, a number of postprocessing options may be considered. For example, in Chapter 8, we report the results from an analysis of isochrony (see Chapter 3 for discussion of isochrony), which can be demonstrated in two ways: (1) when the durations of pen strokes having different lengths do not differ, or (2) when the average velocities of two pen strokes increase in proportion to stroke length. The latter can be reduced to a single score by correlating the average stroke velocity with stroke length. High correlation coefficients indicate velocity scaling and thus adherence to isochrony. In the remaining sections of this chapter, we describe results from kinematic analyses of genuine, forged,[3] and autosimulated or disguised signatures.

Kinematic Approach to Understanding Genuine, Disguised, and Autosimulated Signatures

From observations, FDEs have noted several features that are characteristic of simulated signatures including loss of smoothness or fluency of the writing line, abrupt changes in direction, absence of any regular contrast in pen pressure (point load) between upstrokes and downstrokes, hesitation, unnatural pen lifts, patching, tremor, uncertainty of movement (abrupt changes in direction), and stilted drawn quality handwritings (Osborn 1929; Conway 1959; Harrison 1958; Huber and Headrick 1999; Hilton 1961; Muehlberger 1990; Leung et al. 1993; Alkahtani and Platt 2009). Additionally, Hilton (1961) noted that for a successful forgery, the forger must imitate all habits and qualities of authentic signature and must discard all conflicting elements of his own writing. Harrison (1958) considered the style of signature and difficulty of forgery and noted that "the most difficult to forge is not a florid and practically illegible scrawl, but one which is carefully and accurately written with shaded strokes, and in which each letter can be distinguished."

Traditional forensic research into the static features associated with different signing behaviors has more recently been supplemented by dynamic studies where kinematic data such as pen pressure, stroke formation velocities, and movement durations are collected using digitizing pads in real time. Kinematic analysis techniques have been used to report on the dynamic features associated with disguise and simulation behaviors (van Gemmert et al. 1996). Some researchers are attempting to develop computer algorithms, which can detect disguised handwriting using pattern recognition techniques (De Stefano, Marcelli, and Rendina 2009). Empirical data emerging from both static and dynamic signature research continue to provide FDEs with a resource on which to underpin their opinions based on observations of features in the casework environment.

A study by van Gemmert et al. (1996) provides examples of a neuroscience approach to understanding disguised or forged handwriting. In a study on disguised writing, van Gemmert and colleagues analyzed several kinematic variables (captured using a digitizer tablet and a pressure-sensitive pen). They found that, for stroke size, disguised handwriting was larger and had longer duration than genuine handwriting; however, there were no significant differences in stroke dysfluency. They found pen pressure increased from 1.08 N in genuine to 1.35 N in disguised (free style) samples. Stroke slant was not found to be a discriminatory feature.

Franke (2009) examined kinematic characteristics of signing behavior and found that stroke velocity, pen pressure, and pen lifts or pen stops were not sufficient to discriminate between genuine and forged signatures. The author concluded, "Only the local, inner ink-trace characteristics as well as variations in ink intensity and line quality can provide reliable information in the forensic analysis of signatures."

Van Galen and van Gemmert (1996) looked at the kinematics of genuine and simulated handwriting and found that forgers were successful in copying the spatial aspects of handwriting such as size, slope, and general appearance. However, from the kinematic data, the investigators found that forged handwriting resulted in slower speeds and longer reaction time and was generated by more frequent but smaller force pulses. While pen pressure was higher in simulations, the peak value of pen pressure was higher in the genuine samples. Based on their work, we may conclude that the simulated script is widely different from authentic script, particularly in the kinematic domain (van Galen and van Gemmert 1996). While the van Galen et al. (1996) and the van Galen and van Gemmert (1996) studies were among the first to demonstrate the value of quantitative analyses of pen movement kinematics during handwriting, their findings are based on relatively small sample sizes of a reduced set of kinematic variables. Furthermore, they are based on natural handwriting rather than on signatures.

Current Status of Kinematic Research and Signature Authentication

Previous research has shown that common disguise strategies include changing the formation of capital letters, changing the slant, and changing the speed of writing (Huber and Headrick 1999). The published research does not, however, indicate if writers of mixed, stylized, and text-based signatures employ the same or different disguise strategies. We conducted a series of experiments employing kinematic methods to determine if signing style (text based, mixed, stylized) influences handwriting kinematics equally across the three signature conditions (genuine, disguised, autosimulation).

Methods: Writers and Procedures

The study enrolled 90 subjects (84 right-handed and 6 left-handed writers). Of those who took part in the study, 72% were female; all subjects signed institutionally approved informed consent. Among the subjects, 30 writers naturally wrote text-based signatures, 30 naturally wrote mixed signature styles, and the remaining 30 naturally wrote stylized signatures. This provided a balanced population distribution for writer styles. Each writer was asked to provide 20 signatures (10 genuine, 5 "free-form" disguise, and 5 autosimulation signatures). For the genuine signatures, the subjects were asked to write their normal "check" signature. The free-form disguise and autosimulation scenarios were explained carefully to each participant. Subjects were required to agree verbally that they understood the categories of signatures they would be providing prior to producing the signatures.

It is noted that individuals may normally perform more than one form of genuine signature. For example, a formal signature may be executed on documents such as wills and deeds and a less formal signature may be used for everyday routine transactions. To control for this variable, copies of the same facsimile check were provided to subjects as the sample collection document (see Figure 7.3).

For each signing event, the check was positioned over a Wacom digitizing tablet sampling at 200 samples per second and providing 5μ resolution (Figure 7.4). The tablet was placed on a horizontal table and writers assumed a comfortable writing position while seated (Figure 7.5). The writers were allowed to shift the tablet to assume the angle of writing most comfortable for them. The check was placed in the same position on the tablet for each trial to correct for possible variations in the sensitivity of the tablet surface (Meeks and Kuklinski 1990).

Subjects who took part in pilot tests with the noninking stylus reported that being unable to see an inked line as they wrote was distracting. Since it was not known whether the distraction may have resulted in changes to the normal signing kinematics, an inking stylus was used for all of the signatures collected.

Figure 7.3 Facsimile check provided to subjects to standardize signature collection.

Subjects were asked to write 10 repetitions of their normal signature. These formed the "genuine signatures" (GEN) group. For the free-form disguise, subjects were asked to write five signatures in such a way that they could deny having written them at a later date. They were told to disguise their signature in any way they liked and to use different disguise strategies for each of the five if they wished. The scenario for the disguise was signing a check but the receiver would have no idea of the writer's normal signature style. These

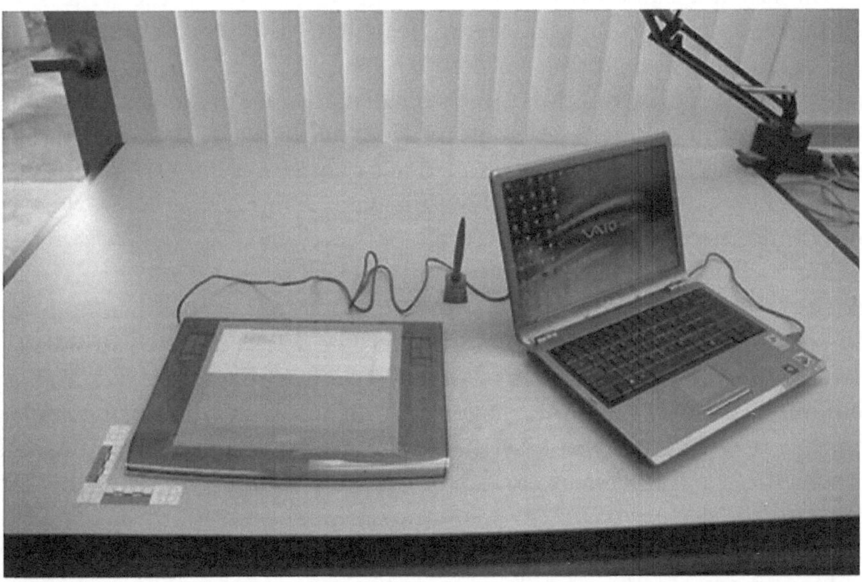

Figure 7.4 Wacom Intuos 3 digitizer tablet and laptop computer. MovAlyzeR software installed on the computer is used to acquire and process signature samples.

Figure 7.5 Subject (to the right) seated in a comfortable writing position while providing samples. The subject could not see the signature on the screen of the laptop as it was recorded.

signatures were referred to as "disguised" (DIS) and intended to represent a situation in which the writer would deny having signed the document.

The last five signatures were also disguised; however, the writers were told that they were to imagine signing each check in a bank where a specimen signature was available for comparison purposes. The signature must therefore be sufficiently similar to their normal signature such that it would likely pass inspection. These signatures were referred to as "autosimulations" (ASIM). The collection process resulted in a database of 1,800 signatures (900 GEN, 450 DIS, and 450 ASIM). Dynamic data from each signature were collected and processed using MovAlyzeR software.

For each segment (i.e., stroke), duration, vertical length, average vertical velocity, average normalized jerk (a measure of pen movement smoothness), and pen pressure were determined. These data were subjected to statistical analyses to evaluate the effects of writer style (text based, mixed, and stylized) and condition (genuine, disguised, and autosimulated) using analyses of variance (ANOVA) and discriminant function analyses. The means for each of these parameters were calculated and compared across the three signature styles (text based, mixed, and stylized) and the three conditions (GEN, DIS, and ASIM). A two-way (style and condition) 3 × 3 ANOVA was used to test any significant differences between the means in the parameters and to look for main effects and interactions. A discriminant function analysis was used to determine if any of the five parameters, or a combination of them, could predict a genuine, disguised, or autosimulated signature.

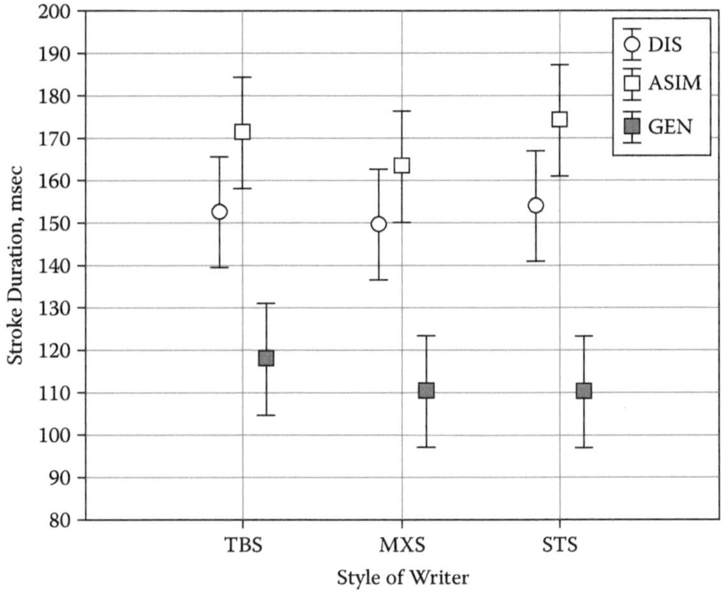

Figure 7.6 Mean (with 95% confidence intervals) stroke duration for three signature conditions (genuine, disguised, and autosimulated) for three groups of writers: text-based (TBS), mixed (MXS), and stylized (STS) forms.

Results

Stroke duration. The results for the comparison of stroke durations for condition (genuine, disguise, and autosimulated) across text-based, mixed, and stylized forms are shown in Figure 7.6. We found a significant effect of condition ($F2,261 = 57.67$; $p < 0.001$). Genuine signatures were found to have less duration (were written more quickly) than both types of disguised signatures. However, the effect of writer style was not statistically significant ($F2,261 = 0.74$; $p > 0.10$), nor was there a significant condition by style interaction.

Stroke length. The results for vertical stroke length are shown in Figure 7.7. We found a significant main effect for writer style ($F2,261 = 15.43$; $p < 0.001$) and condition ($F2,261 = 15.76$; $p < 0.001$). The text-based signatures were found to be smaller than the mixed and stylized signatures ($p < 0.001$), whereas mixed and stylized signatures did not differ significantly in size. Genuine signatures were larger than both mixed and stylized signatures with no significant difference between the latter two. A significant interaction was found between style and condition ($F4,261 = 5.72$; $p < 0.001$); however, the effects of this interaction vary. Although there was a style effect for genuine signatures, no such effect of writer style was found for either disguised or autosimulated signatures.

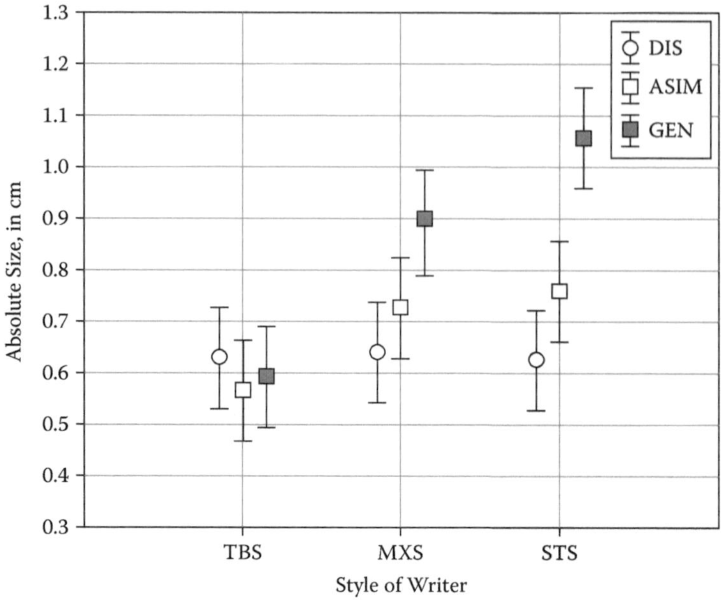

Figure 7.7 Mean (with 95% confidence intervals) stroke length for three signature conditions (genuine, disguised, and autosimulated) for three groups of writers: text-based (TBS), mixed (MXS), and stylized (STS) forms.

Stroke velocity. The results for average vertical stroke velocity are shown in Figure 7.8. As with stroke length, we found a significant effect for writer style ($F2,261 = 22.14$; $p < 0.001$) and condition ($F2,261 = 45.19$; $p < 0.001$). Text-based signatures are written more slowly than mixed or stylized signatures. No significant differences were found between mixed and stylized signatures when compared across the three conditions. Stroke velocities for disguised and autosimulated signatures were not found to differ. A significant interaction was also found between style and condition ($F4,261 = 8.56$; $p < 0.001$). The effect of style was significant for genuine signatures but not for disguised or autosimulated signatures.

Normalized jerk. The results for average normalized jerk are shown in Figure 7.9. We found a significant main effect for condition ($F2,261 = 12.01$; $p < 0.01$). Genuine signatures displayed less jerk (written more fluently) than disguised and autosimulated signatures. However, there were no effects for writer style ($F2,261 = 2.39$; $p > 0.10$), nor was there a condition by style interaction ($F4,261 = 0.99$; $p > 0.10$).

Pen pressure. The results for pen pressure are shown in Figure 7.10. We found a significant main effect for writer style ($F2,261 = 6.46$; $p < 0.01$) and condition ($F2,261 = 4.18$; $p < 0.01$). Text-based signatures were written with less pen pressure than mixed or stylized signatures while the latter two styles were similar. Genuine signatures were written with greater

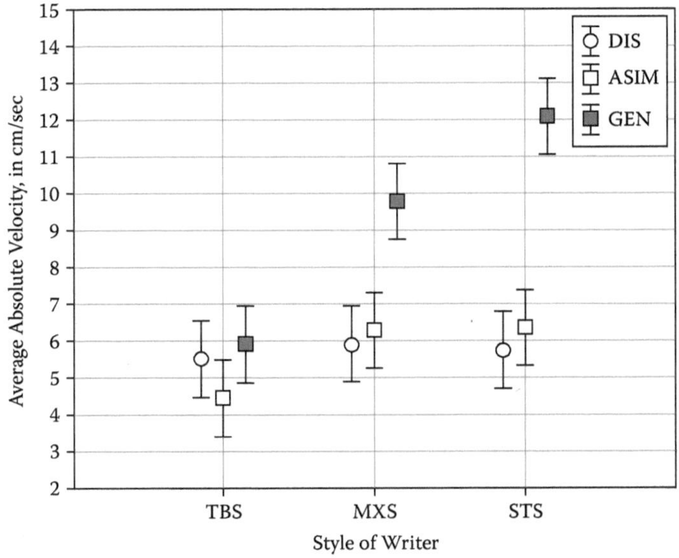

Figure 7.8 Mean (with 95% confidence intervals) stroke velocity for three signature conditions (genuine, disguised, and autosimulated) for three groups of writers: text-based (TBS), mixed (MXS), and stylized (STS) forms.

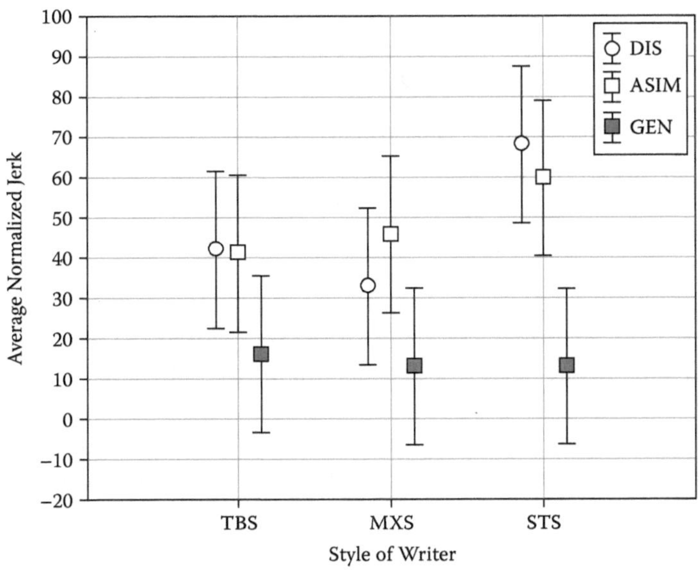

Figure 7.9 Mean (with 95% confidence intervals) average normalized jerk (smoothness) for three signature conditions (genuine, disguised, and autosimulated) for three groups of writers: text-based (TBS), mixed (MXS), and stylized (STS) forms.

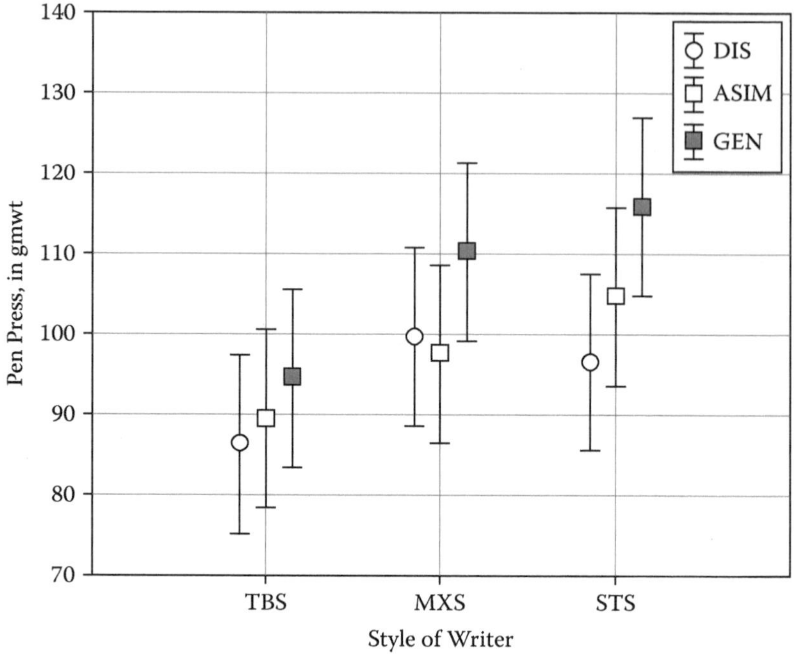

Figure 7.10 Mean (with 95% confidence intervals) pen pressure/stroke for three signature conditions (genuine, disguised, and autosimulated) for three groups of writers: text-based (TBS), mixed (MXS), and stylized (STS) forms.

average pen pressure than autosimulated or disguised signatures. We did not observe an interaction between style and condition for the measure of pen pressure ($F_{4,261} = 0.40$; $p > 0.10$).

Results from the discriminant function analysis testing whether groups of writers could be distinguished based on measured parameters are shown in Table 7.2.

Using a five-parameter kinematic model, genuine signatures were distinguished from disguised and autosimulated signatures with greater than 80% accuracy. However, the model was unable to distinguish disguised from autosimulated signatures: Accuracy was less than 70%. Attempts to improve the classification of disguised and autosimulated signatures using a two-factor model (consisting of stroke size and velocity) netted an increase in accuracy to only 71.6%.

Research by FDEs has shown that writers disguise their signatures by changing the slant, shape, size, speed, and fluency of letters. However, previously reported research did not attempt to discriminate between different styles of signatures.

The present study investigated whether there is any relationship between signature styles and the conditions of genuineness, disguise, and

Table 7.2 Results of a Discriminant Function Analysis Showing Accuracy (%) for Classifying Signature Condition for Different Writer Styles

Comparison	Condition predicted	Writer style		
		Text	Mixed	Stylized
GEN versus DIS	Genuine	90.0	83.3	76.6
	Disguised	76.6	83.3	90.0
	All	83.3	83.3	83.3
DIS versus ASIM	Disguised	63.3	73.3	70.0
	Autosimulated	70.0	60.0	63.3
	All	66.6	66.6	66.6
GEN versus ASIM	Genuine	90.0	90.0	93.3
	Autosimulated	80.0	73.3	80.0
	All	85.0	81.6	86.6

autosimulation. We hypothesized that handwriting kinematics would differ across conditions and that these differences would vary as a function of style. We found that some, but not all, parameters differed between the different signature styles. Specifically, for text-based signatures, duration was an important discriminator between genuine and both disguised and autosimulated signatures. However, the disguised and autosimulated signatures could not be separated by the duration parameter. Therefore, if FDEs could reliably determine the duration of a text-based signature from a static trace, the rate of accuracy of determinations whether such signatures are genuine or disguised/autosimulated could be increased.

For mixed-style signatures, velocity and size were found to be significant in separating genuine from both disguised and autosimulated signatures. Genuine and autosimulated signatures could be distinguished by considering their duration. Lastly, for stylized signatures, three parameters—velocity, size, and jerk (dysfluency)—were significant in separating genuine from both disguised conditions, while duration was important in separating genuine from autosimulated signatures. This indicates that FDEs have a better chance of discriminating between genuine and both disguised conditions if signatures are stylized rather than text based or mixed.

Genuine signatures were written with more pen pressure than disguised and autosimulated signatures. It might be expected that a writer would apply more pressure when disguising his or her signature because more thought is required in executing the disguise. Van Gemmert et al. (1996) found that "increase of pen pressure is higher in the cursive than in the printing style samples of disguised script," which "may seem as a confirmation of the view that using a disguised print letter style is less demanding than using a cursive style." Writers who utilize printed letter forms as a disguise are apt to exert less pen pressure in their disguised signatures.

It was interesting that no significant difference was noted in pen pressure for the genuine and disguised signatures. This is in agreement with previous studies that found that "generally speaking, the overall pressure patterns of a writer's signature have been shown to be habitual and highly individualistic to that writer" (Estabrooks 2000) and that "dynamic pressure patterns are an integral part of an individual's signature" (Tytell 1998). If pen pressure is an ingrained motor-control characteristic, then even though a writer is disguising his or her signature, this writing habit may be too powerful to change. Further research is needed to confirm this finding.

Summary

The aim of this chapter was to present a quantitative approach to the dynamic analysis of handwriting and signatures. While the vast majority of research regarding signatures has focused on static traces, modern technology has enabled researchers to quantify the kinematic features of signatures at the level of an individual pen stroke. Historically, visually detectable features in handwritten signatures formed the basis of evidence supporting whether a questioned signature is genuine, disguised, or forged. Research into static features associated with different signing behaviors can be supplemented by dynamic studies where kinematic data are collected from writers signing on digitizing tablets. This technique has been used to describe the kinematic characteristics of disguise and simulation behaviors in terms of pen pressure, stroke formation, and movement duration.

Our research on the differences in kinematic features between genuine, disguised, and autosimulated signatures provides strong empirical support for the notion that stroke size, speed, and fluency are important factors in differentiating genuine signatures from disguised signatures. The results also underscore the importance of the style of the specimen signature when evaluating whether a questioned signature is genuine, disguised, or autosimulated. An ongoing challenge among FDEs is the development of reliable methods to measure objective parameters from static signatures quantitatively. Some work has been done in this regard by researchers attempting to breach an automatic signature recognition system using dynamic features recovered from a static signature using software (Hennebert et al. 2007). These investigators achieved useful results by regaining the velocity and pressure profiles of the genuine static signature.

Modern kinematic approaches that utilize digitizing tablets combined with sophisticated software can be very powerful tools in collecting dynamic signature and handwriting data. The resulting databases can then be statistically analyzed to determine interactions between writing styles and writing conditions.

This information will provide FDEs with empirical data that will assist them in their evaluations of kinematic information from static signatures.

Notes

1. Wacom (www.wacom.com).
2. Neuroscript, LLC, Tempe, AZ (www.neuroscriptsoftware.com).
3. Simulated signatures are more commonly referred to as forgeries. However, forgery is a legal term and for the purposes of this chapter, the term *simulation* will be used.

Isochrony in Genuine, Autosimulated, and Forged Signatures

8

Introduction

In forensic signature examination, the examiner has to determine whether a signature is authentic or forged. There is no shortage of contemporary research into the ability of forensic document examiners (FDEs) to express valid opinions on the authorship of questioned signatures (Found, Sita, and Rogers 1999; Kam et al. 2001; Sita, Found, and Rogers 2002). In general, the problem of differentiating between a disguised signature and a forged signature stems from observations that writers repeatedly choose the same strategies when disguising their signatures (Herkt 1986; Mohammed 1993; Wendt 2000; Durina 2005). Found and Rogers (2009) concluded that FDEs were better at determining whether signatures were simulated than they were at identifying genuine signatures and were more likely to make an error when judging a genuine signature to be a simulation than by judging a simulated signature to be genuine.

Most writers attempting to disguise their signature impart obvious changes in their handwriting, such as different letter shapes, angle of slant, etc., while in many instances retaining attributes of the finer structure of the signature. In some cases, however, it may be very difficult or even impossible to discriminate between authentic but disguised signatures and forgeries. Herkt (1986) found that FDEs reported disguise to be difficult and also concluded that it is difficult at times to discriminate between an autosimulation and a forgery.

We reasoned that measuring a fundamental component of the handwriting motor program such as isochrony may be an effective approach to differentiating between autosimulations and forgeries. If successful, this method would provide the FDE with a quantifiable means of judging whether a signature was genuine, autosimulated, or even forged, thus improving reliability.

Minimization principles are applied across scientific disciplines such as physics (mechanics), evolutionary biology, and engineering (see Chapter 3). In biology, minimization principles describe a means by which animals attempt to achieve maximum effectiveness with minimum effort when executing a goal-directed movement. A fundamental compensatory mechanism called isochrony can be applied to many areas of motor control (Viviani and Terzoulo 1980). In the broadest qualitative terms, the principle states that the velocity of voluntary movements increases with the extent of the movement, thus keeping execution time approximately constant (Viviani and McCollum 1993). This principle is observed in a wide variety of motor behaviors (Vinter and Mounoud 1991). Adherence to isochrony has been observed with small movements, but the relationship fails with larger movements (Grossberg and Paine 2000).

The isochrony principle holds that the average velocity with which the gesture (writing and drawing) is executed increases spontaneously as a function of its amplitude, so execution time is less dependent on size than it would be otherwise (Viviani and Terzoulo 2008). This reduces the demand on the motor program by minimizing the need to include a temporal parameter. Thus, movements of varying extent (such as different heights of handwritten letters) can be preprogrammed using fewer variables.

If handwriting movements adhere to principles of minimization of effort and are programmed to ensure efficiency, then one would expect nonprogrammed movements, such as forgeries or autosimulations, not to adhere to this principle. That is, a forged signature is not likely to be learned or produced with kinematic efficiency. As such, forged handwriting movements are less likely to display properties associated with highly programmed movement.

We conducted a study of genuine, forged, and autosimulated signatures to investigate whether autosimulated and forged signatures adhere to the principle of isochrony and whether genuine signatures can be distinguished from forged signatures on the basis of an isochrony analysis. We hypothesized that handwriting kinematics for normal genuine signatures will adhere to the isochrony principle, whereas forgeries or autosimulations will not.

Methods: Writers and Procedures

We conducted two experiments to test the hypothesis that genuine signatures would differ from forged or autosimulated signatures on the basis of a single measure of isochrony. In the first experiment, 60 writers were asked to write their own signatures 10 times and to forge three model signatures 15 times each. Among the 60 writers, there were 20 in each of three signature style groups: text based, stylized, and mixed. In the second experiment, 90 writers were asked to write their own signatures 10 times and to simulate (or disguise) their signature 5 times each. Among the 90 writers from experiment 2, there were 30 in each of three signature style groups: text based, stylized, and mixed.

All signatures were digitized using a Wacom digitizing tablet and MovalyzeR software. Data collection, summarization, and analyses were performed using procedures described in Chapter 7. For the purpose of these experiments, the analysis of forged (experiment 1) and autosimulated (experiment 2) signatures involved only those signatures that matched the style of the writer.

Vertical stroke size and velocity features were extracted from each pen stroke of an individual's signature, regardless of the style of writing. Movalyzer software was used to extract the amplitude and average velocity of each pen stroke for the entire signature. The number of strokes varied with signature length. For each signature, we then calculated the correlation coefficient for the relationship between stroke size and average stroke velocity using Pearson r-procedures. As an index of the amplitude–velocity relationship, strong correlation coefficients (i.e., $r > 0.80$) would suggest adherence to the isochrony principle.

A two-way analysis of variance (ANOVA) in a mixed design was performed to test whether the mean correlation coefficient for genuine signatures differed from the mean coefficient for autosimulations for a given writer style (mixed, text based, or stylized). Writer style with three levels served as the between-group factor, and signature type with two levels (genuine or autosimulated) served as the within-subject factor.

Results

Table 8.1 shows the mean correlation coefficients for the linear relationship between stroke length and average stroke velocity for genuine versus forged signatures (experiment 1) and genuine versus autosimulated signatures (experiment 2).

Experiment 1: genuine versus forged. There was a significant difference in the correlation between absolute size and average absolute velocity for genuine signatures as compared to forgeries for all three styles of signatures. These results are shown in Figure 8.1.

Table 8.1 Mean Correlation Coefficients for the Experiments on Isochrony

	Experiment 1 (n = 30 in each group)		Experiment 2 (n = 20 in each group)	
	Genuine	Forged	Genuine	Autosimulated
Text based	0.84 (0.09)	0.74 (0.13)[a]	0.84 (0.07)	0.66 (0.10)[a]
Mixed	0.81 (0.11)	0.76 (0.12)[a]	0.80 (0.06)	0.66 (0.10)[b]
Stylized	0.79 (0.13)	0.75 (0.09)[a]	0.78 (0.13)	0.54 (0.17)[b]

[a] $p < 0.0001$.
[b] $p < 0.05$.

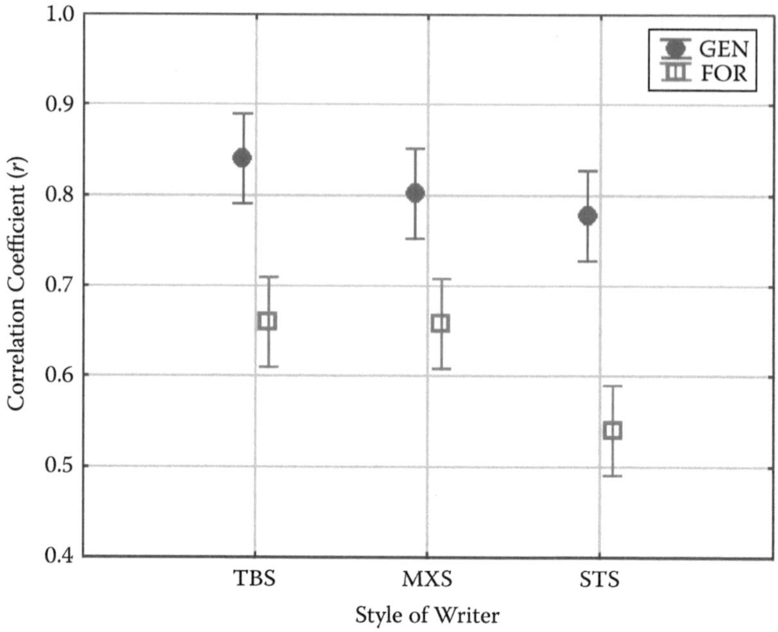

Figure 8.1 Mean correlation coefficients for the relationship between absolute stroke size and absolute stroke velocity for genuine and forged signatures for three groups of writers.

Experiment 2: genuine versus autosimulated. For text-based writers (TBS), there was a significant difference in the correlation between absolute size and average absolute velocity for genuine signatures as compared to autosimulations. There was no significant difference for mixed (MXS) or stylized (STS) writers for the genuine and autosimulated signatures. These results are shown in Figure 8.2.

Discussion

Several new findings emerged from this study. First, kinematic analyses of stroke size revealed that the movements forming genuine signatures adhered to the principle of isochrony. This is based on the presence of a strong linear relationship between stroke amplitude and average stroke velocity. For autosimulations, the size–velocity relationships were noticeably weaker, although not significantly different from the writers' genuine signatures for mixed or stylized. Text-based writers, on the other hand, exhibited significantly lower size–velocity correlation coefficients for forged and autosimulated signatures compared to their genuine signatures.

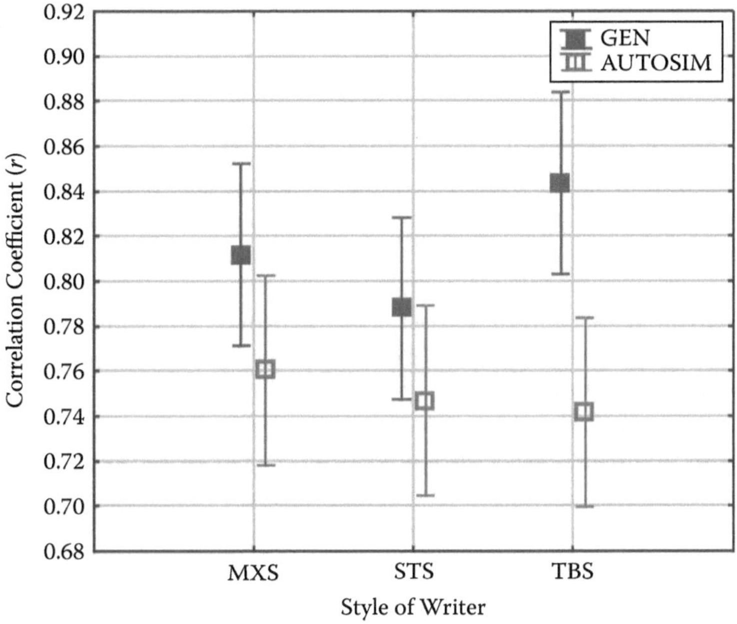

Figure 8.2 Mean correlation coefficients for the relationship between absolute stroke size and absolute stroke velocity for genuine signatures and autosimulations for the three groups of writers.

Analyses of the features associated with forged signatures (experiment 1) revealed weak relationships between stroke size and velocity, suggesting nonadherence to the isochrony principle. These results suggest that while genuine signatures adhere to the isochrony principle, forged signatures do not. As may be expected, autosimulations show minimal adherence to isochrony.

A second finding of the present study was that adherence to the isochrony principle varied across writer subgroups. Specifically, stylized writers showed no difference in mean coefficient correlation for absolute size and absolute velocity between their genuine signatures and forged signatures. Text-based writers, on the other hand, showed significant differences between their genuine signatures and their forgeries. Mixed-style writers (a combination of text-based and stylized forms) demonstrated variable adherence.

When genuine signatures were compared with forged signatures, there was a significant difference in the size–velocity correlation coefficients for all signature styles. Since genuine signatures are the product of open loop control and forged signatures generally are the product of closed loop control, these results are supportive of the findings based on children's handwriting by Vinter and Mounoud (1991). They found that the handwriting of 5- to 6-year-old children conformed to the isochrony principle. When the child turns 7 and begins to apply more feedback to his or her handwriting with regard to the length of strokes, less isochrony is associated with the movement outcomes. As the child becomes a more experienced writer and reduces reliance on feedback (8 years old), isochrony is again in evidence (Vinter and Mounoud 1991).

It is clear that any condition that affects the normal progress of an abstract motor program may well impact whether isochrony is preserved. In the production of a forged signature, the forger is trying to capture the habits perceived in the model signature and simultaneously attempting to discard his own writing habits. In terms of motor control, the forger's program for executing signatures will be altered. This is displayed in the clear differences between genuine and forged signatures for correlation between absolute size and absolute velocity.

The subjects in this study all provided their genuine signatures, autosimulations, and forgeries at one sitting. It would be expected that the variation in these signatures would be more limited than if they had provided the signatures over a period of time. For the forgeries, the subjects were allowed three practices before they provided the 15 forgery attempts. With more time to practice, some of them may have become better at forging and their forgeries may have shown greater adherence to the isochrony principle.

In experiment 1, the forgers were under no pressure to produce a very good forgery, which may not reflect a real-life situation. There was no penalty for the subject if the forgeries were not adequate. The subjects were not compensated for their effort. However, previous research has shown that this may not impact subjects' motivation (Kam, Fielding, and Conn 1998).

The subjects in both experiments 1 and 2 were healthy as far as could be determined and the forgeries were of signatures that were produced by healthy writers. If other circumstances that may affect the ability of the subjects to write, such as illness, age, medication, or alcohol, were introduced, this could affect the results. It has been shown previously, for example, that diseases such as Parkinson's could affect isochrony in handwriting (van Gemmert, Adler, and Stelmach 2003).

The subjects in these experiments were laypeople with no experience of forensic handwriting examination or specialized training in handwriting. Further experimentation with expert penmen such as trained calligraphers (Dewhurst, Found, and Rogers 2009) may provide useful data.

Summary

The isochrony principle holds that the average velocity with which the gesture is executed increases spontaneously as a function of the extent of movement, so execution time is less dependent on movement extent than it would be otherwise. This reduces the demand on the motor program by minimizing the need to include a temporal parameter. Thus, movements of varying extent (such as different heights of handwritten letters) can be programmed using fewer parameters. In the present study, we reasoned that if genuine natural handwriting movements adhere to principles of minimization of effort and are programmed to ensure efficiency, then one would expect non-programmed movements, such as forgeries or autosimulations, to violate these principles. That is, a forged signature is not likely to be programmed or produced with kinematic efficiency. We hypothesized that handwriting kinematics for normal genuine signatures will adhere to the isochrony principle, whereas forgeries or autosimulations will not.

Our results support this hypothesis. Thus, while genuine signatures adhered to the isochrony principle, forged signatures did not. For autosimulations, the stroke length–velocity relationship was significant for text-based writers only.

We may conclude that alterations in the execution of a motor program contribute to inefficient movements that can be detected by observing adherence to the isochrony principle. When producing a forged signature, the forger is trying to capture the habits perceived in the model signature and simultaneously attempting to discard his or her own writing habits. That is, the forger is attempting to overwrite his or her own motor program for handwriting. When this occurs, movements appear inefficient and violate the isochrony principle.

Kinematic Analyses of Stroke Direction in Genuine and Forged Signatures

9

Introduction

The study of kinematics is mainly concerned with motion characteristics of a subject and examines this from a spatial and temporal perspective without reference to the forces causing the motion. Kinematic analyses provide descriptions of movement to determine how fast an object moves and how high or how far it travels. As a result, position, velocity, and acceleration are of particular interest in kinematics. Thus, data can be derived from any anatomical structure having any starting position (or joint angle) moving to any defined endpoint (Godfrey et al. 2008).

A handwriting pattern may be viewed as a sequence of ballistic strokes (Teulings and Schomaker 1993) comprising a series of upstrokes and downstrokes, which may or may not be concatenated. Upstrokes and downstrokes influence the way in which handwriting is perceived. Maarse and Thomassen (1983) found that the slant of handwriting is determined by the downstrokes and noted that downstrokes appear to be more stable than upstrokes.

Van Galen and Weber (1998) studied the kinematics of handwriting upstrokes and downstrokes as a means to understand the nature of the handwriting motor program. Their aim was to see if the motor program was composed of discrete and integral sets of movement goals or whether the program was more generalized, reflecting a segmented sequence of goal trajectories. In their study, 12 subjects were instructed to write a series of nonsense words. Pen movements were recorded on a digitizing tablet. During the writing task, the horizontal writing space was either kept the same or unexpectedly extended or shortened by 7%. The investigators found that vertical stroke amplitude adapted to these spatial constraints as they occurred, with upstrokes showing more of an adaptation than downstrokes.

Their finding is consistent with the notion that downstrokes are more stable than upstrokes in terms of vertical size (Maarse and Thomassen 1983; Teulings and Schomaker 1993). Teulings and Schomaker comment that downstrokes seem to be the information carriers of handwriting. Historically, forensic document examiners have noted that, in simulations, evidence of tremor is seen more often in curved upstrokes than downstrokes. Osborn (1929) noted:

> The connecting upward strokes are especially significant for the comparison of movement impulses, as these strokes show the propulsive power of the writer. In slow or unskillful writing the upward strokes, or some of them at least, are usually produced with more smoothness and freedom than downward strokes, and just the opposite condition is usually found in fraudulent writing.

In this chapter we report the results of an experiment to test whether kinematic measures of upstrokes and downstrokes distinguish forged signatures from genuine signatures. Prior literature suggests several differences in the kinematics of upstrokes versus downstrokes only for genuine signatures. Our goal was to extend these observations to the study of forged signatures. We hypothesized that the forged signatures would exhibit a different pattern from genuine signatures, which can be detected from the kinematic relationships between upstrokes and downstrokes.

Methods: Writers and Procedures

The study enrolled 60 writers, 20 of whom naturally wrote text-based signatures, 20 who naturally wrote mixed signature styles, and the remaining 20 who naturally wrote stylized signatures. This provided a balanced population distribution for writer styles. Each of 20 writers was asked to provide 10 genuine signatures and to forge a similarly styled signature 15 times. For their genuine signatures, subjects were asked to write their normal "check" signature. To assist in the accuracy of the forged signatures, subjects were given a model signature to replicate. It is noted that individuals may normally perform more than one form of genuine signature. For example, a formal signature may be executed on documents such as wills and deeds and a less formal signature may be used for everyday routine transactions. To control for this variable, copies of the same facsimile check were provided to subjects as the sample collection document shown in Figure 7.3 in Chapter 7.

For each signature trial, the facsimile check was positioned over a Wacom digitizing pad sampling at 200 samples per second and providing 5μ resolution. The tablet was placed on a horizontal table and writers assumed a comfortable writing position while seated. The writers were allowed to shift the tablet to assume the angle of writing most comfortable for them. The check was placed in the same position on the tablet for each trial to correct for possible variations in the sensitivity of the tablet surface (Meeks and Kuklinski 1990).

Dynamic data from each signature were collected using MovAlyzeR software. For each stroke, we calculated duration (in milliseconds), vertical length (in centimeters), average vertical velocity (in centimeters per second), average normalized jerk (a measure of pen movement smoothness), and pen pressure (in arbitrary units). Kinematic data for the upstrokes and downstrokes were coded on the basis of the directional sign assigned to the average velocity for a given stroke. Thus, for all strokes having negative velocities, the associated duration, vertical lengths, average velocities, average normalized jerk scores, and pen pressures were coded separately from the kinematic parameters associated with strokes having positive velocities. For each writer, the average value for each of the five kinematic scores was calculated for upstrokes and downstrokes for the genuine and forged conditions, yielding a total of 240 scores for each kinematic parameter.

A mixed-model analysis of variance (ANOVA) was used to test simple main effects of writer style and condition (genuine versus forged) on the difference in stroke direction for each kinematic parameter. For each ANOVA, writer style (three levels) and condition (two levels) served as between-group factors, while stroke direction (two levels) served as the within-subject factor. In addition, we calculated an upstroke/downstroke difference score and used t-tests to evaluate difference between genuine and forged conditions for each writer group.

Results

Table 9.1 shows the means (and standard deviations) for the five kinematic parameters for three writer groups for genuine and forged signatures. For stroke duration, the ANOVA revealed a significant main effect for condition ($F(1,114) = 103.41$; $p < 0.0001$), with genuine signatures having significantly shorter stroke duration than forged signatures, and a significant effect of stroke direction ($F(1,114) = 6.35$; $p < 0.05$). For all writer styles, stroke duration was shorter for genuine than forged signatures. Downstrokes had shorter durations than upstrokes for forged signatures (regardless of writer style); for genuine signatures, there was no discernable difference in duration between upstrokes and downstrokes.

For stroke length, the ANOVA revealed a significant main effect for writing style ($F(2,114) = 26.49$; $p < 0.0001$) with text-based writers exhibiting shorter stroke lengths than stylized or mixed writers. We found a significant main effect for condition ($F(1,114) = 20.93$; $p < 0.0001$); genuine signatures had significantly longer stroke lengths than forged signatures, and there was a significant effect of stroke direction ($F(1,114) = 237.52$; $p < 0.000001$). For all writer styles, stroke length was shorter for forged than genuine signatures and downstrokes were shorter than upstrokes.

Table 9.1 Means (Standard Deviations) for the Five Kinematic Parameters for Three Writer Groups for Genuine and Forged Signatures

Writer Group	Condition	Stroke direction	Duration (ms)	Length (cm)	Velocity (cm/s)	ANJ	Pressure
Text							
	Genuine	Up	109 (14)	0.64 (0.27)	6.76 (2.86)	13 (5)	475 (175)
		Down	109 (15)	0.56 (0.25)	6.12 (2.59)	13 (5)	551 (171)
	Forged	Up	254 (121)	0.51 (0.14)	3.39 (0.89)	344 (591)	357 (134)
		Down	239 (122)	0.47 (0.13)	3.26 (0.82)	343 (589)	412 (141)
Mixed							
	Genuine	Up	120 (22)	0.81 (0.23)	8.87 (2.71)	18 (13)	418 (149)
		Down	117 (28)	0.69 (0.21)	7.81 (2.39)	18 (14)	500 (172)
	Forged	Up	278 (109)	0.67 (0.09)	4.23 (1.28)	213 (208)	271 (225)
		Down	263 (100)	0.62 (0.09)	3.91 (1.21)	213 (207)	327 (235)
Stylized							
	Genuine	Up	113 (25)	1.39 (0.65)	15.41 (7.20)	16 (16)	481 (148)
		Down	111 (21)	1.20 (0.61)	13.56 (6.85)	16 (15)	542 (152)
	Forged	Up	338 (171)	0.96 (0.20)	5.23 (1.97)	224 (198)	367 (151)
		Down	316 (141)	0.57 (0.15)	4.06 (1.63)	224 (197)	366 (153)

Findings for average stroke velocity were similar to those for stroke length. The ANOVA revealed a significant main effect for writing style (F(2,114) = 20.71; p < 0.0001); text-based writers exhibited lower average stroke velocities than stylized or mixed writers did. We found a significant main effect for condition (F(1,114) = 88.04; p < 0.00001)—genuine signatures had significantly longer stroke lengths than forged signatures did (9.16 cm/s versus 3.74 cm/s)—and a significant effect of stroke direction (F(1,114) = 87.45; p < 0.00001). In general, average stroke velocity was faster for genuine than forged signatures and downstrokes were slower than upstrokes.

Mean upstroke/downstroke difference scores for stroke velocity are shown in Figure 9.1 for three writer groups for genuine and forged signatures. The figure shows that stylized writers reduce velocity for forged signatures (as expected) but, unlike text-based (TBS) and mixed (MXS) writers, stylized (STS) writers do not show a difference in the upstroke/downstroke difference score for forgeries. Other writers show significantly less difference in stroke direction for forgeries than for genuine samples.

For our measure of smoothness (average normalized jerk, ANJ), the ANOVA revealed a significant main effect for condition (F(1,114) = 25.01; p < 0.0001) with forged signatures having significantly less smoothness (higher ANJ scores) than genuine signatures. We found no significant effect of stroke

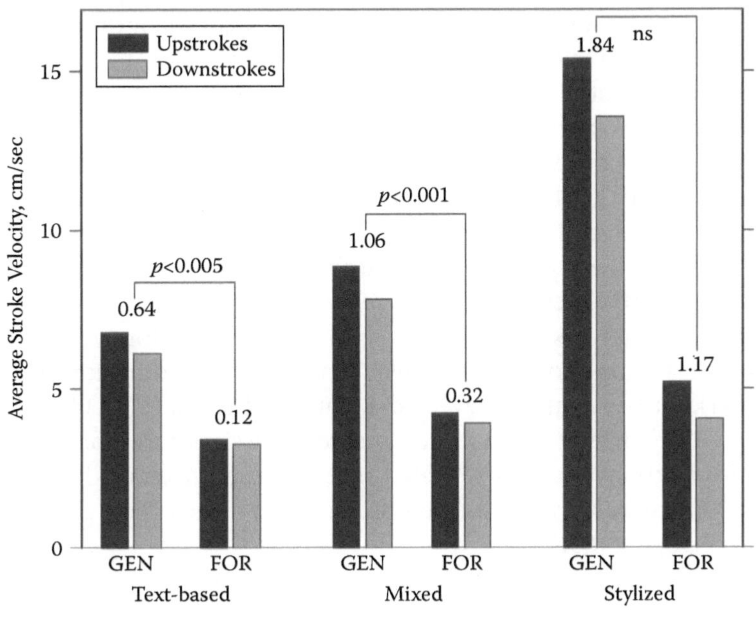

Figure 9.1 Mean upstroke/downstroke difference scores for average stroke velocity for three writer groups for genuine and forged signatures. The numbers above each pair are the difference scores. Lines (with p-values) are for t-tests applied to the genuine versus forged difference. ns = nonsignificant.

direction on our measure of smoothness. Thus, for all writer styles, forged signatures were less smooth than genuine signatures with no discernable difference in smoothness between upstrokes and downstrokes.

For pen pressure, the ANOVA revealed a significant main effect for condition ($F(1,114) = 22.17$; $p < 0.0001$); forged signatures had significantly less pen pressure than genuine signatures. We also found a significant effect of stroke direction ($F(1,114) = 115.56$; $p < 0.00001$). Thus, for all writer styles, forged signatures were written with less pen pressure than genuine signatures and, with the possible exception of forged signatures by stylized writers, downstrokes were associated with greater pen pressure than upstrokes were.

Upstroke/Downstroke Ratio Scores

Table 9.2 shows the upstroke/downstroke ratios for each kinematic variable for the three writer groups for genuine and forged signatures. Ratios greater than 1.00 indicate greater values for the upstroke than downstroke for a given kinematic parameter.

A significant effect of condition was found for the stroke length ratio ($F(1,114) = 13.27$; $p < 0.001$) with higher ratios for forged (mean = 1.30) than genuine (mean = 1.19) signatures. Interestingly, there was significant interaction between writer group and condition for the stroke length ratio. Whereas text-based and mixed writers exhibited greater ratios for genuine than forged signatures, stylized writers exhibited the opposite effect, with greater upstroke/downstroke ratios for forged (1.72) than genuine (1.20) signatures.

While writer groups differed in their upstroke/downstroke ratios for stroke velocity ($F(2,114) = 11.20$; $p < 0.0001$), there were no effects of condition on the ratio overall. However, further analyses of the velocity ratio indicated a trend ($t = 1.83$; $p = 0.07$) for a difference between forged and genuine signatures for stroke velocity for stylized writers only.

Statistical analyses revealed no significant effects for writer group or condition in the upstroke/downstroke ratio for stroke duration or smoothness (ANJ). Lastly, analyses revealed a statistical trend ($F(1,114) = 2.93$; $p = 0.08$) for a difference between forged and genuine signatures for the pen pressure ratio. Further analyses revealed that stylized writers exhibited higher upstroke/downstroke ratios for pen pressure for forged than genuine signatures ($t = 3.30$; $p < 0.01$).

Summary

The present study contributes to the understanding of important differences in the production of genuine versus forged signatures. The findings supported

Table 9.2 Means (Standard Deviations) Upstroke/Downstroke Ratios for Each Kinematic Variable for the Three Writer Groups for Genuine and Forged Signatures

Writer Group	Condition	Duration (ms)	Length (cm)	Velocity (cm/s)	ANJ	Pressure
Text						
	Genuine	1.00 (0.09)	1.17 (0.15)	1.12 (0.12)	1.00 (0.00)	0.85 (0.11)
	Forged	1.08 (0.18)	1.09 (0.08)	1.04 (0.12)	1.00 (0.00)	0.87 (0.10)
Mixed						
	Genuine	1.05 (0.18)	1.20 (0.19)	1.15 (0.15)	1.00 (0.00)	0.84 (0.11)
	Forged	1.06 (0.17)	1.09 (0.06)	1.09 (0.12)	1.00 (0.00)	0.81 (0.14)
Stylized						
	Genuine	1.02 (0.15)	1.20 (0.24)	1.18 (0.25)	1.00 (0.02)	0.89 (0.14)
	Forged	1.06 (0.19)	1.72 (0.20)	1.31 (0.16)	1.00 (0.01)	1.01 (0.08)
All						
	Genuine	1.02 (0.15)	1.19 (0.20)	1.15 (0.18)	1.00 (0.01)	0.86 (0.12)
	Forged	1.06 (0.18)	1.30 (0.32)	1.15 (0.18)	0.99 (0.01)	0.89 (0.13)

previous literature showing differences between upstrokes and downstrokes for genuine signatures along several kinematic parameters, including stroke length (19% longer for upstrokes), stroke velocity (15% higher for upstrokes), and pen pressure (14% lower for upstrokes) across writer styles. The study revealed new findings on differentiating forged from genuine signatures based on analysis of upstroke/downstroke ratios.

Specifically, we found that the ratio for stroke length was significantly greater in forged than genuine signatures for stylized writers, but lower in forged signatures for text-based or mixed writers. For stroke velocity, we observed increase in the ratio (from 18% to 31% greater velocity for upstrokes) from genuine to forged signatures for stylized writers. Lastly, we found that stylized writers exhibited lower pen pressures for upstrokes than downstrokes (11%) for forged signatures, which was not observed for genuine signatures. For all other writer groups, we observed consistently lower pen pressures for upstrokes than downstrokes for both genuine and forged signatures.

Unlike previous studies demonstrating *dynamic* kinematic differences between forged and genuine signatures written by the same writer (see Chapters 7 and 8), the present findings include kinematic features that can also be quantified using *static* handwritten samples. A forensic document examiner using existing tools can evaluate stroke length and pen pressure from known and questioned historical documents for judgments of authenticity. Our findings suggest that accurate measures of stroke length and calculating the upstroke/downstroke ratio or difference can increase the scientific validity and reliability of judgments of authenticity.

Neurologic Disease, Drugs, and the Effects of Aging

III

Neurological Disease and Handwriting

10

Introduction

Over 40 years ago, Hilton (1969) described a subset of troubling signatures that examiners of questioned documents often encounter. These signatures are those produced during serious illness. The problem is amplified if the illness occurs late in life, when handwriting changes from natural causes related to aging are already underway. Recognizing that variation is present in nearly every identifiable element of a signature, Hilton stressed the importance of obtaining multiple samples to appreciate the variability and, more importantly, to identify a pattern of change falling outside this range of normal variability during periods of illness.

Consistent with understanding of the neurobiological mechanisms that control handwriting from nearly half a century ago, Hilton noted that the extent to which a signature may be affected by serious illness depended on how the illness affected the writer's coordination and physical strength. Hilton reasoned that signatures written during serious illness lack the consistency of the person's earlier handwriting and contain elements that vary in an "erratic manner…[and] may contain a number of extraneous false strokes" (p. 160). In cases of severe illness, Hilton observed that "letter design, ratio between tall and short letters, slant, and alignment of the signature to ruled lines all lack stability" (p. 160). While devoid of scientific rigor, these observations nonetheless are relevant today as they point to the direction of needed research to clarify disease-driven sources of variation in handwriting.

Much of the contemporary research on handwriting in disease conditions is descriptive in nature and aims to associate deteriorating handwriting with progressive disease (Huber and Headrick 1999). Unlike gait, posture, strength, speed, and coordination, changes in handwriting may not signal the onset of a progressive neurological disease. The one exception may be micrographia associated with Parkinson's disease (PD). Approximately 10% of the patients diagnosed with PD exhibit micrographia a few years prior to onset of other parkinsonian signs (McLennan, Nakano, and Tyler 1972). An interesting case in support of the diagnostic importance of micrographia in PD is the case of Sir Henry Head (1861–1940), the famous British neurologist of the early 1900s. Henry Head was diagnosed with PD in 1919; however, documents uncovered and reported by Pearce (2008) showed that Sir Henry's signatures exhibited undeniable micrographia 13 years prior to the diagnosis.

Diagnostic specificity refers to the ability of a behavioral or laboratory test to classify an individual accurately as having a specific disease or condition versus some other condition. The diagnostic specificity of deviant handwriting is not strong. While deteriorating handwriting is ubiquitous among the many neurological and psychiatric illnesses, it lacks sufficient specificity to serve as a useful marker of a specific disease. The vast majority of the correlative research linking disease with deteriorating handwriting is based on qualitative methodologies and subject to examiner bias and questionable reliability. For example, while micrographia is a well recognized feature of handwriting in PD (Wing 1980; Margolin and Wing 1983; Margolin 1984; Tucha et al. 2006), not all PD patients exhibit micrographia (McLennan et al. 1972; Tarver 1988). In a large-scale prevalence study, McLennan et al. (1972) estimated that 10%–15% of PD patients exhibit micrographia. Insofar as tremor is ubiquitous in advanced age, neurological disease, and fatigue, ascribing tremulous handwriting to a specific disease is problematic.

Osborn (1929) cautioned that slight tremor observable in handwriting provides limited causal information. Throughout the published literature, attempts by forensic document examiners to associate deviant handwriting to a particular neurological condition have been largely unsuccessful (e.g., Boisseau, Chamberland, and Gauthier 1987; Tarver 1988; Willard 1997). Boisseau et al. (1987) reported that many of their patients with clinically apparent PD or ET exhibited relatively intact handwriting. While some of their PD patients exhibited improvement in handwriting form following levodopa therapy, others did not. The authors failed to detect a pattern of handwriting impairment associated with a specific disorder. It is important to remember that these conclusions are based on qualitative, subjective analyses of static handwriting and lack the necessary sensitivity, precision, and reliability necessary to address this question. As demonstrated later in this Chapter, modern quantitative approaches to the study of the dynamic kinematic features of handwriting are more likely to reveal important characteristics that can discriminate one disease process from another, thus providing the document examiner with stronger case arguments.

However, certain features of handwriting can help narrow the range of possible diagnoses based on what we know about how the brain controls handwriting (see Chapter 1) and what we know about the relationship between a specific neurological disease and causal neural mechanisms. For example, weakness leading to an inability to grip the writing instrument and slow pen movements yielding low pen pressures against the writing surface are more likely to be associated with lower motor neuron disease (such as amyotrophic lateral sclerosis [ALS]) than an extrapyramidal disease (such as PD). Tremulous handwriting movements can occur in PD, essential tremor (ET), progressive supranuclear palsy (PSP), and Huntington's disease (HD)—all involving the extrapyramidal system—but rarely in peripheral neuropathy or ALS. Dysfluent, jerky handwriting characterized by random intrusive strokes is common in HD and rare in patients with cortical strokes. Overlapping pathophysiology in the form of shared circuits and neurochemistry weakens the ability to ascribe deficient features observed in handwriting to a particular cause or neurological condition.

Nonetheless, a strong case may be built for ascribing micrographia to a specific neurobiological mechanism. Micrographia develops in conditions that affect movement scaling, such as PD (Wing 1980; Margolin and Wing 1983; Margolin 1984). Observations of handwriting from individuals with a variety of conditions that alter the normal functions of the basal ganglia suggest that micrographia is not unique to PD (Gilmour and Bradford 1987; Barbarulo et al. 2007). Rather, micrographia appears to be a consequence of processes that interrupt striatal dopamine neurotransmission.

Observations of handwriting movements among individuals with neurological disease can inform underlying pathological mechanisms responsible for the disease and can provide a record of change in disease progress or benefits of treatment. In the following sections, we summarize prior published work and recent research from our laboratory on handwriting kinematics associated with common neurological disorders including PD, ET, HD, PSP, and AD (Alzheimer's disease). While time course and clinical management differ for these conditions, there is some overlap in their neurochemistry and pathophysiology, particularly with regard to subcortical brain regions that govern motor control. Given the overlapping brain regions thought to be involved in the expression of motor problems in PD, ET, HD, PSP, and AD, it is reasonable to hypothesize that these conditions would also show overlapping patterns of abnormal handwriting kinematics.

Handwriting in Specific Neurological Diseases

Parkinson's Disease

Hallmark motor signs of PD include tremor at rest, muscular rigidity, bradykinesia, and postural imbalance. Parkinsonian resting tremor is a slow, coarse tremor generally observed in the upper extremities with arms at rest or during ambulation. Unlike essential tremor (see later discussion), parkinsonian tremor amplitude decreases and tremor may stop completely during voluntary action, such as handwriting. In practice, muscular rigidity is observed as resistance to passive movement or stiffness. Clinicians evaluating muscular rigidity generally rotate the patient's hand, arm, leg, or head and assess resistance; it is not uncommon for tremor to embed itself during this maneuver. This form of rigidity is called "cogwheel rigidity" to reflect the ratcheting nature of resistance the limb has through the arc of movement.

Parkinsonian rigidity impairs mobility, limits the extent of movement, and can contribute to painful muscular cramping, particularly during periods of sustained handgrip posture as in handwriting. Bradykinesia, or slowing of movement, leads to myriad functional impairments, including hygiene, eating, dressing, speech, walking, and handwriting. Control of fine movements is particularly compromised. The term bradykinesia refers to slowing; however, the simple act of producing a pen stroke actually involves at least three different elements, each of which is impaired in PD. They include akinesia (loss or poverty of movement), hypokinesia (reduced movement extent), and bradykinesia (slowness of movement). In evaluating handwriting in PD, it is important to keep these three aspects separate as they each contribute to the overall impairment in different ways and to varying degrees.

In their review of the relevant literature through the 1990s, Huber and Headrick (1999) noted that handwriting difficulties might be one of the earliest indications of PD. Reductions in speed (bradykinesia) and size of handwriting (micrographia or hypokinesia) are not uncommon early in the course of the disease, while sequencing, completeness, and linguistic aspects are preserved in handwriting. Walton (1997) noted that handwriting from her PD patients tended to show similar but exaggerated changes to those reported in otherwise healthy older writers (see Chapter 13). Micrographia was observed predominantly among the PD patients with earlier onsets of disease, occurring in about 17% of these patients. Other features reported by Walton in her comprehensive assessment of PD handwriting include greater number of pen lifts (about twice that of age-comparable healthy subjects) and greater variability in stroke length both within and between samples, particularly among late-onset PD patients.

Micrographia and Bradykinesia

A recurrent theme in any discussion of parkinsonian handwriting is the nature of micrographia, which was initially described by Lewitt (1983). Individuals with micrographia are unable to sustain their normal size letter formation for more than a few seconds (Teulings and Stelmach 1991). Early in the disease, there is a progressive decrease in letter height within a sample sentence or signature. Later in the course of the disease, diminished letter or stroke size is present at the onset of the writing sample. McLennan et al. (1972) reported no statistical association between micrographia and presence of any other hallmark motor sign in their study of 95 PD patients.

While the prevalence of micrographia is relatively low, it has potential as a marker of disease progress within an individual. Specifically, once it has been determined that micrographia exists, its severity is likely to progress along with natural disease progression. Seemingly random patterns of micrographia in longitudinal handwriting samples may actually reflect the pattern of medication dosing or the fluctuating pattern often seen as pharmacotherapeutic effects diminish. To the informed examiner, micrographic handwriting patterns may help validate self-reported histories of disease variability and its pharmacological management.

In an attempt to expand our understanding of the nature and variability of motor impairment in PD, researchers have employed sophisticated methods to quantify specific kinematic elements of handwriting. These studies have consistently demonstrated PD results in impairment of multiple kinematic aspects of handwriting, including vertical stroke size, speed, acceleration, peak pen pressure, and stroke duration (Teulings and Stelmach 1991; Muller and Stelmach 1992; Flash et al. 1992; Longstaff et al. 2001; van Gemmert, Teulings, and Stelmach 2001; Tucha et al. 2006). In the following sections we summarize this research as well as recent observations from our laboratory.

Two important studies focusing on the effects of levodopa therapy on handwriting kinematics shed light on the nature of the handwriting disturbance in PD. Tucha et al. (2006) and Lange et al. (2006) examined the effects of levodopa therapy on handwriting kinematics in a group of 27 and 12 PD patients, respectively. For both studies, patients were asked to produce handwritten combinations of words and letters containing the letter sequence "*ll*" on a digitizing tablet. Analyses were performed to extract kinematic data from ascending and descending strokes. Patients were studied off and on their usual dopaminergic medications.

The researchers found that, off medication, PD subjects produced the "*ll*" sequence with longer stroke durations with lower amplitude, velocity, and acceleration compared with healthy subjects. These group differences were independent of medication status; however, with the exception of total writing time, handwriting kinematic scores improved for the PD patients on medication. Because there was overlap in the patients reported in the Tucha et al. (2006) and the Lange et al. (2006) papers, their findings are essentially indistinguishable. Figure 10.1 shows traces from a single healthy subject and PD patient, on and off medication, depicting reductions in vertical stroke amplitude, velocity, and acceleration typically observed in PD.

The precise mechanism responsible for micrographia in PD is not known. Dounskaia et al. (2009) hypothesized that one potential mechanism may stem from the failure to make use of the many degrees of freedom and flexibility associated with multijoint coordinated movements, as is the case with handwriting. The researchers designed an interesting study to test whether

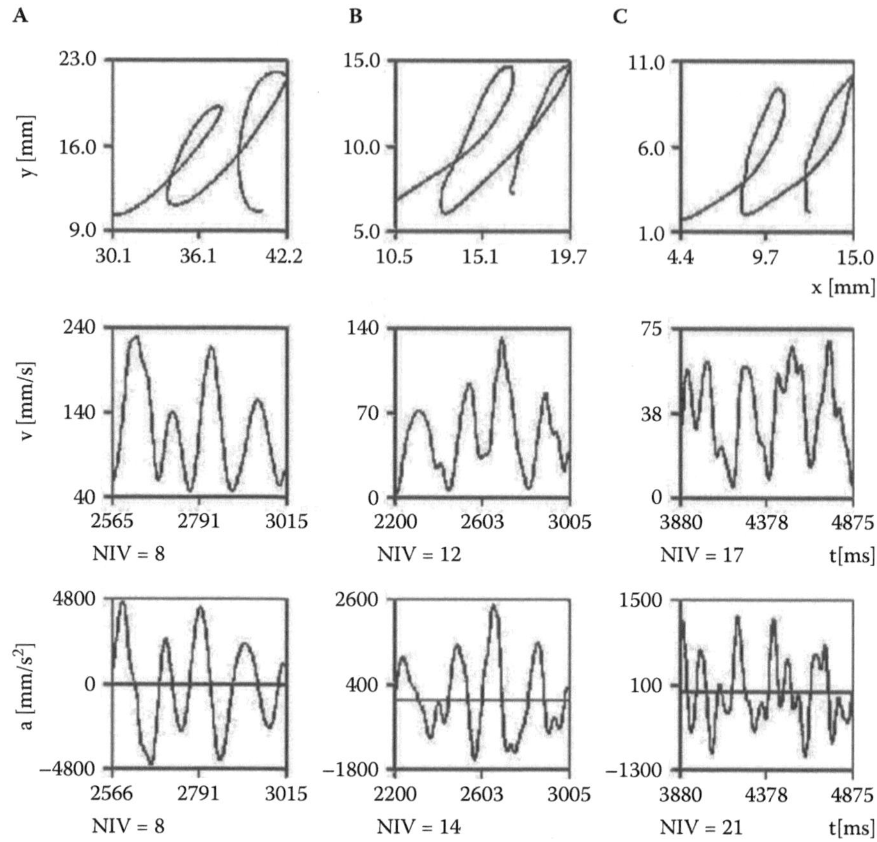

Figure 10.1 Handwriting specimens of the letter combinations *"ll"* with corresponding velocity and acceleration profiles of a healthy subject (A), a PD patient on dopaminergic medication (B), and the same PD patient off dopaminergic medication (C). NIV = number of inversions. (From Tucha, O. et al. 2006. *Journal of Neural Transmission* 113:609–623. With permission.)

manipulating the position of the wrist and fingers (ranging from partially extended to flexed) in various combinations (degrees of freedom) led to predictable changes in the form of various handwritten shapes. Nine PD and nine age-comparable healthy comparison subjects were studied.

The authors found that for nearly all experimental manipulations, PD patients resorted to a preferred set of kinematic parameters failing to utilize the full range of parameters offered by the experimental manipulations. In other words, PD patients could not coordinate finger and wrist movements across the range of positions imposed by the experiment. Based on their findings, the authors proposed that micrographia in PD may stem from an inability to incorporate multiple degrees of freedom associated with multijoint coordinated movement into their handwriting repertoire. In this sense, micrographia may be a manifestation of restricting the degrees of freedom necessary to produce the full range of letter shapes and sizes available to the healthy writer.

Handwriting Dysfluency in PD

Dysfluent movements are common in PD and are thought to stem from one of two distinct underlying mechanisms. The first is tremor, which, as noted before, is a resting tremor in PD. However, as the disease progresses, tremor becomes more pervasive and is present during posture and action. Attempts to suppress tremulous handwriting can appear as dysfluency. The second mechanism is related to chronic exposure to dopaminergic medications. As PD progresses, cells within the substantia nigra continue to die off, leaving relatively few projections capable of transmitting dopamine from the substantia nigra to the striatum. It is thought that as the dopaminergic neurons are depleted, there is a corresponding reduction in the number of striatal receptors (as fewer are needed).

Unlike the healthy brain, which manages to balance the amount of neurotransmitter (in this case, dopamine) with the number of receptors, the parkinsonian brain cannot regulate this balance. The imbalance caused by too much replacement dopamine and too few dopamine receptors creates a state of *hyperdopaminergia.* Whereas *hypodopaminergia* leads to slowness and reduced movement, hyperdopaminergia leads to excessive movement, dyskinesia, jerkiness, and loss of smoothness. In the later stages of PD, this phenomenon is known as levodopa-induced dyskinesia with hourly fluctuations referred to as "on-off" phenomona.

Kinematically, there are several ways to quantify handwriting dysfluency. One common method is to count the number of times the velocity or acceleration profile changes directions (inversion). Smooth handwriting movements occur with a constant velocity. Dysfluent handwriting is characterized by abrupt changes (or inversions) in velocity. Tucha et al. (2006) examined handwriting fluency in their PD patients by calculating the number of velocity and acceleration inversions. They reported a greater number of velocity and acceleration inversions among PD patients vis-à-vis healthy comparison subjects. Furthermore, they reported an increase in the number of inversions when patients were off rather than on medication.

However, to interpret these and other findings of handwriting change in PD, it is important to consider the timing of the handwriting assessment relative to medication dosing. PD patients with levodopa-induced dyskinesia generally develop dyskinetic hand movements within 45 minutes of taking levodopa (Caligiuri and Lohr 1993). This time delay corresponds to the time required for levodopa plasma level to reach its peak. As Tucha et al. did not report the time interval between medication dosing and handwriting assessment, it is questionable that the study was even designed to assess the levodopa mechanism. Furthermore, assessments for the off medication state were performed at least

15 hours following the last medication dose, so it is likely that their patients were somewhat symptomatic at the time when handwriting was assessed.

An intriguing experiment conducted by Lange et al. (2006) strengthens the importance of dopamine neurotransmission in the maintenance of handwriting fluency. This study involved only the healthy subjects from their prior experiments. Subjects were grouped into those with normal and those with abnormal findings from transcranial sonography showing hyperechogenic activity in the subcortical brain region approximating the location of the substantia nigra. Berge et al. (1999) had previously demonstrated that individuals with sonographically verified substantia nigra hyperechogenicity exhibited a reduced dopamine uptake in the striatum. Lange et al. (2006) reported that their group of healthy participants with substantia nigra hyperechogenicity exhibited significantly more velocity and acceleration inversions, confirming that handwriting dysfluency is sensitive to alterations in nigrostriatal neurotransmission.

Velocity Scaling and Isochrony in PD Handwriting

Several studies have shown that when PD patients write loops or spirals, they produce stroke velocities that are independent of stroke amplitudes (van Gemmert, Adler, and Stelmach 2003; Viviani et al. 2009). That is, PD patients do not scale stroke velocity appropriately for a given stroke amplitude as well as healthy comparison subjects do. Van Gemmert et al. (2003) reported that unmedicated PD subjects were impaired relative to medicated PD patients and controls in their ability to scale peak acceleration with increasing stroke size during drawing of outward spirals. Teulings et al. (1991) and van Gemmert et al. (1999) showed that while medicated PD subjects undershoot pen movement distances when instructed to increase the stroke height, their movement times were normal. This also suggests that PD subjects fail to increase movement velocity in order to attain the proper movement amplitude while maintaining normal temporal control.

The inability to scale velocity or acceleration to accommodate changes in stroke amplitude suggests that the writer fails to adhere to the isochrony principle (as reviewed in Chapter 3). Specifically, producing uniform stroke velocities across varying stroke amplitudes can only be accomplished by increasing stroke duration, which violates the isochrony principle. Failure to adhere to this important minimization principle suggests that the handwriting disturbance in PD stems from a disturbance in motor programming.

The isochrony disturbance is common in PD for a wide range of movements. Studies of single-joint wrist rotation showed that patients with parkinsonian bradykinesia did not scale movement velocity properly with increasing movement amplitude (Caligiuri et al. 1998; Pfann et al. 2001; Robichaud et al. 2002). Patients with more severe motor signs, particularly bradykinesia, exhibited lower velocities and lower velocity scaling (VS) scores.

Using data extracted from a previous study on handwriting in PD (Caligiuri et al. 2006), we evaluated whether PD patients adhere to the isochrony principle during natural handwriting. Study subjects consisted of 13 individuals (nine males and four females) diagnosed with idiopathic Parkinson's disease and 12 (10 males and two females) normal, healthy control subjects (NC). The mean (±sd) duration since their initial diagnosis of PD was 8.9 (5.5) years. The mean ages for the PD and NC subjects were 66.7 (9.6), and 53.0 (12.9) years, respectively. All PD subjects were treated with some form of dopamine-replacement therapy at the time of study.

Prior to subjects' undergoing the handwriting task, we rated the severity of parkinsonism using the unified Parkinson's disease rating scale (UPDRS; Fahn and Elton 1987). Based on the motor exam portion of the UPDRS (part 3), 6 of the 13 PD patients were considered to have moderate motor impairment and 7 were rated as mild or minimal at the time of study. The means (±sd) on the UPDRS motor subscale for the six PD patients with moderate and seven with mild motor impairment were 28.3 (±3.8) and 13.4 (±3.1), respectively; higher scores reflected greater motor impairment.

Handwriting kinematics were quantified using a noninking pen on a Wacom digitizing tablet (30 × 22.5 cm, sampling rate 120 samples per second, RMS accuracy 0.01 cm) attached to a notebook computer running MovAlyzeR software. Subjects were instructed to write the word "hello" twice from left to right and to stay within the upper and lower boundary lines drawn on a piece of white paper with their dominant (right) hand. Three conditions were administered: boundary line heights of 1, 2, and 4 cm. Figure 10.2 shows a writing pattern produced by a healthy control subject. We used the MovAlyzeR software to filter, segment into up and down strokes, and extract the kinematic and temporal features for each stroke.

Data reduction consisted of determining vertical stroke length (in centimeters) and peak velocities (in centimeters per second) for the medial "*ll*" segments in the cursive writing pattern "hello" for each amplitude condition. Subsequently, the slope of the linear regression of the vertical peak velocity versus stroke height was calculated. The slope coefficient (in centimeters per second per centimeter) served as the measure of velocity scaling (VS).

The results are shown in Table 10.1. Of the 13 PD subjects, 12 had velocity-scaling coefficients below the 95th percentile of the NC mean (3.35 cm/s/cm). Several significant correlations between the pen movement variables and independent measures of symptom severity were significant for the PD subjects. Specifically, lower VS slope coefficients correlated with higher total scores from part 3 (motor exam) of the UPDRS ($r = -0.65$; $p < 0.05$), especially when using only hand movement speed of the UPDRS ($r = -0.81$; $p < 0.01$) (see Figure 10.3). There was only a marginally negative correlation with finger tapping speed (from the UPDRS) ($r = -0.53$; $p < 0.10$). Also, peak velocity for the 4 cm stroke height condition was negatively

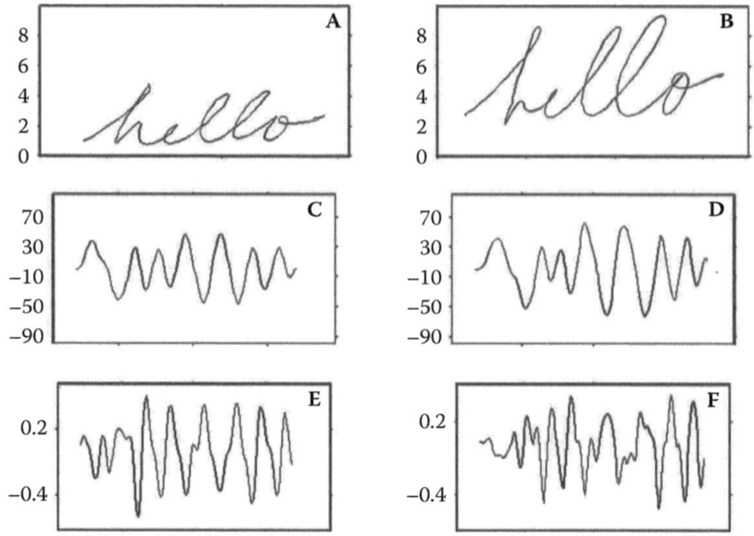

Figure 10.2 Handwriting samples of size of the word "hello" written by a healthy subject: lower boundary trials (left) and higher boundary trials (right). Vertical size (in cm) is shown in A and B; vertical velocity (in cm/s) in C and D; and vertical jerk (in cm/s^3) in E and F. The criterion size was verified by measuring the vertical size of the middle *"ll"* segments in A and B. The mean peak velocity was measured by average of the (absolute) velocity peaks in C and D. The number of positive and negative jerk peaks was counted in E and F.

Table 10.1 Means (Standard Deviations) for Peak Vertical Stroke Velocity for the Medial *"ll"* Segments of the Word "Hello" Written within Three Vertical Boundary Heights and the Velocity Scaling Score

	NC (n = 12)	PD (n = 13)	Statistic
Velocity 1 cm, cm/s	9.22 (1.64)	7.05 (3.79)	1.83 ($p > 0.10$)
Velocity 2 cm, cm/s	14.71 (3.73)	11.55 (4.73)	1.84 ($p > 0.10$)
Velocity 4 cm, cm/s	26.69 (6.90)	18.61 (6.63)	2.92 ($p < 0.01$)
VS slope coefficient, cm/s/cm	5.69 (1.91)	3.81 (1.61)	2.67 ($p < 0.05$)

Notes: NC = normal healthy subjects; PD = patients. The statistical t-scores (and p values) are for the NC versus PD difference.

correlated with the total motor score of the UPDRS ($r = -0.70$; $p < 0.02$) and, specifically, with the bradykinesia score (i.e., hand movement speed) ($r = -0.62$; $p < 0.05$). Advanced age in the PD subjects was significantly related to lower VS slope coefficients for PD ($r = -0.87$; $p < 0.01$); older PD subjects exhibited lower VS scores than older patients. Age was not related to VS score for the healthy subjects in this study.

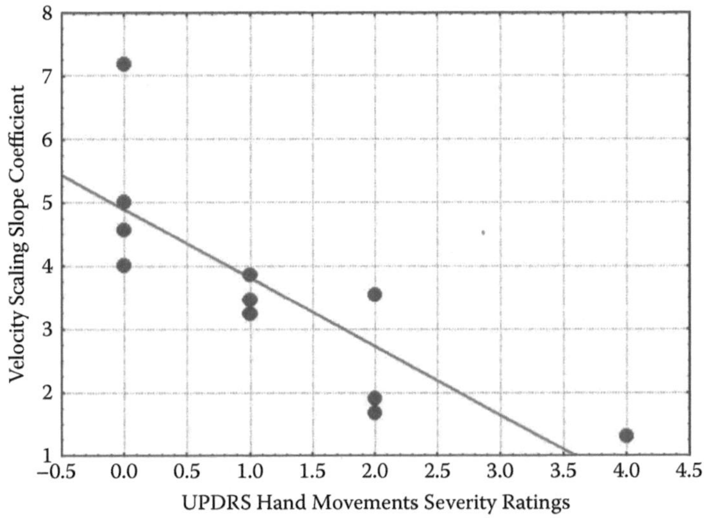

Figure 10.3 Scatterplot showing the relationship between velocity scaling coefficient for handwriting and clinical severity of bradykinesia (based on rapid alternating hand movements) in patients with PD.

The results were subjected to discriminant function analyses to identify sensitivity and specificity of the VS score for classifying subjects as parkinsonian or healthy based on handwriting kinematics. Results indicated that the VS slope coefficient had 90% sensitivity and 60% specificity. That is, solely on the basis of the VS score derived from handwritten samples of the word "hello," 90% of the PD subjects were correctly classified as PD. However, the VS score could not accurately classify a healthy writer as healthy.

Our findings are consistent with previous literature on handwriting and limb movements in PD. Specifically, PD patients appeared to increase movement duration to compensate for an inability to increase velocity when greater movement distances are needed. Given that the motor program appears to code for uniform movement durations and increasing movement velocity when increasing movement extent, these findings support the idea that parkinsonian bradykinesia likely stems from a disturbance in the programming, storage, or retrieval of motor elements to assure cost minimization.

Effects of Deep Brain Stimulation on Handwriting Kinematics in a PD Patient

Deep brain stimulation (DBS) is a common surgical procedure used to treat a variety of disabling neurological symptoms including tremor, rigidity, stiffness, bradykinesia, and gait problems. At present, DBS is FDA approved for patients whose symptoms cannot be sufficiently controlled with medications.

DBS uses a surgically implanted, battery-operated neurostimulator—a small device about the size of a stopwatch. The DBS system consists of an electrode, an extension wire, and the neurostimulator unit. The extension and electrode are surgically inserted through a small opening in the skull and implanted in the brain. The tip of the electrode is positioned within the targeted brain area. When turned on by the surgeon or patient, the neurostimulator delivers electrical stimulation to selected targeted areas in the brain that control movement. The electrical stimulation interrupts the abnormal neurotransmission that produces tremor and other PD motor signs. Common targets in DBS are the thalamus, subthalamic nucleus, or globus pallidus, depending on the profile and severity of parkinsonian motor signs.

We examined handwriting kinematics from a single PD patient who had recently received a DBS implant. The stimulating electrode terminals were placed bilaterally in the subthalamic nuclei. The subject was a 59-year-old male with PD of moderate severity. He had a 14-year history of PD and a 4-year history with the DBS. With the DBS unit turned off, his total score on the motor subtest of the UPDRS was 43, reflecting moderate motor impairment. With the DBS unit switched on, his UPDRS motor score improved to 26, reflecting clinical improvement; however, some motor impairment remained.

We recorded handwriting movements during two sessions: one with the stimulator turned on and another with the stimulator turned off. Handwriting movements were recorded and analyzed using MovAlyzeR software under standard procedures described elsewhere in this and previous chapters. Samples were acquired using a Wacom digitizing tablet and inkless pen. The patient was instructed to produce continuous sequences of the cursive letter "*l*," alternating cursive "*lleellee*," overlay circles repeated at a normal writing speed and again as fast as possible (each at least eight times), and the sentence "Today is a nice day" using cursive script with the dominant hand (for a total of five tasks). The patient was instructed to stay within a marked 2 cm vertical boundary when performing the handwriting tasks. Five trials were administered for each task.

Using MovAlyzeR software, we extracted several kinematic variables from each ascending and descending stroke within and across trials. Four kinematic variables were of interest in this case as they reflect important parkinsonian motor features. These included stroke duration, vertical stroke length (micrographia), vertical stroke velocity (bradykinesia), and average normalized jerk (dyskinesia) per stroke. In addition, we assessed velocity scaling (adherence to the isochrony principle) by calculating the correlation coefficients for the relationship between vertical stroke length (size) and vertical velocity for the "*lleellee*" sequences. This task was selected for the isochrony analysis because it consists of a wide range of the stroke length and velocity values necessary for statistical analyses. High correlations ($r > 0.80$) indicate that as vertical stroke length increases, there is a proportional increase in stroke velocity reflecting normal velocity scaling and thus adherence to the isochrony principle.

Figure 10.4 shows handwriting samples for the patient under two conditions: DBS unit turned off and DBS unit turned on. Shown are the raw handwriting samples for the word "Today" from the sentence "Today is a nice day" (A and B) and the corresponding velocity traces (C and D). The exemplary traces show that with the DBS unit turned off, there is a reduction in the range of velocity peak excursion. The raw waveform and velocity trace show prolonged segments during which movement seems to have stopped or been replaced by low-amplitude tremor (at the 1 s mark in the velocity trace). In contrast, with the DBS unit turned on, movements are more fluid, as exemplified by the continuously oscillating velocity trace. Once the handwriting movement is underway, the average peak velocity of movement is markedly higher for the "on" than "off" condition.

Table 10.2 shows the means for the five kinematic variables for each handwriting condition. Results from this single case clearly demonstrate that activating the DBS unit imparts significant improvement in the kinematic features of handwriting. For three of the five tasks, stroke velocity increased significantly when recorded with the DBS unit switched from off to on. For repetitive movements such as continuous letter "*lllll*" or overlay circles, stroke length increased when the DBS unit was turned on, suggesting a reduction in parkinsonian micrographia. Interestingly, pen movements were significantly smoother with the DBS unit turned on than when it was in the off position on four of the five tasks, including sentence production.

Figure 10.5 shows the effects of DBS status on peak stroke velocity across the five handwriting tasks. It can be seen that under conditions of optimal stimulation with the DBS turned on, while movement velocity is increased compared to the off position, the benefit was not observed for all handwriting tasks. The greatest effect was found for the task requiring the patient to write overlay circles as rapidly as possible (overlay F); however, no effects of DBS were observed for sentence writing or alternating production of cursive "*llee*." This suggests that DBS may have limited effects on movement speed for complex alternating handwriting sequences.

Figure 10.6 shows the results of our analysis of velocity scaling (isochrony). The top scatterplot shows the relationship between stroke length and velocity for pen movements during cursive writing of alternating sequences of the letters "*lleellee*" with the DBS unit turned off. The stroke length and its corresponding velocity for each vertical stroke for all trials were included in this analysis. The scatterplots suggest a modest relationship between stroke length and velocity such that, as stroke length increased, velocity increased. However, there were many strokes that appeared to violate this relationship. The bottom scatterplot shows this same relationship during cursive writing of the same alternating letter sequence with the DBS unit turned on. It is

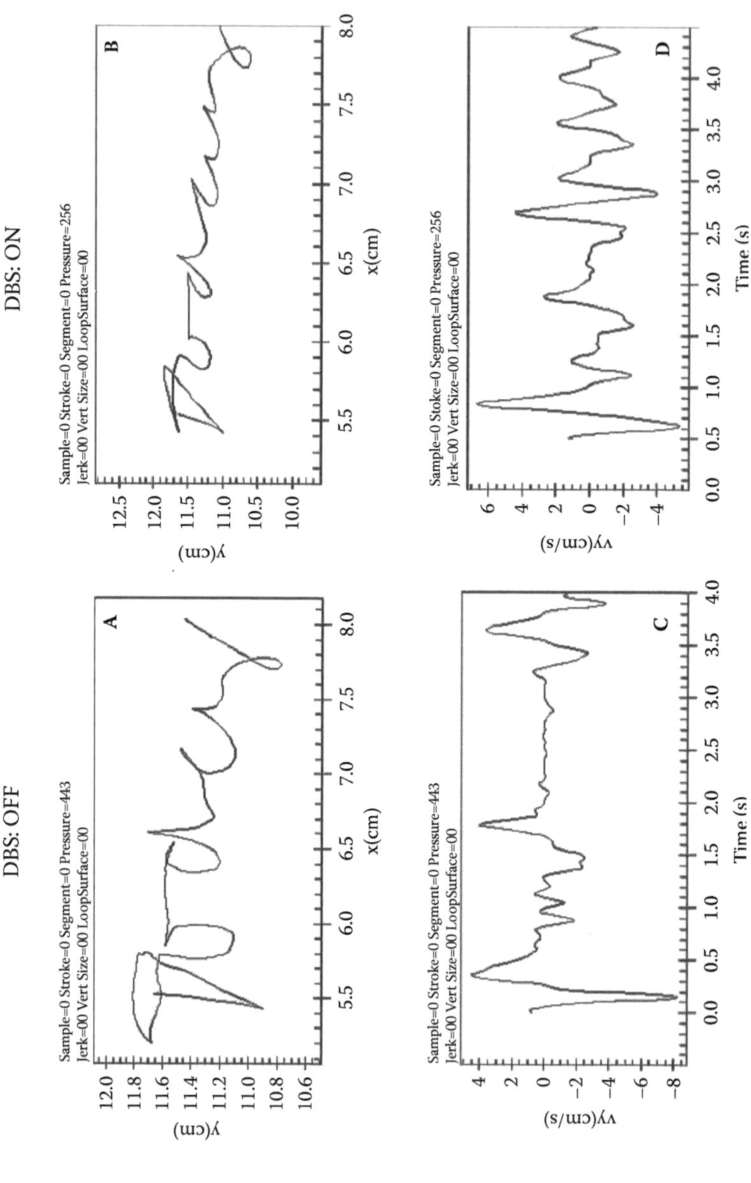

Figure 10.4 Handwriting samples for a single PD patient under two conditions: DBS unit turned off (A and C) and DBS unit turned on (B and D). Shown are the raw handwriting samples for the word "Today" from the sentence "Today is a nice day" (A and B) and the corresponding velocity traces (C and D).

Table 10.2 Mean Kinematic Scores for Five Stroke Variables for Five Handwriting Tasks from a Single PD Patient Recorded with the DBS Unit Turned Off and On

Task	DBS Status	Stroke Duration (ms)	Stroke Length (cm)	Peak Stroke Velocity (cm/s)	Average Normalized Jerk
lllll	Off	610	0.83	2.67	98.4
	On	636	1.07[b]	3.44[b]	102.9
lleellee	Off	610	0.77	2.55	143.3
	On	493[a]	0.73	2.82	44.5[b]
Sentence	Off	249	0.34	2.26	39.2
	On	198	0.38	2.47	19.1[b]
Overlay circles: normal	Off	569	0.89	2.83	62.2
	On	472[b]	1.04[b]	3.72[b]	29.1[b]
Overlay circles: fast	Off	469	0.96	2.93	30.6
	On	265[b]	0.97	5.12[b]	13.8[b]

[a] Significantly different from off condition at $p < 0.001$.
[b] Significantly different from off condition at $p < 0.0001$.

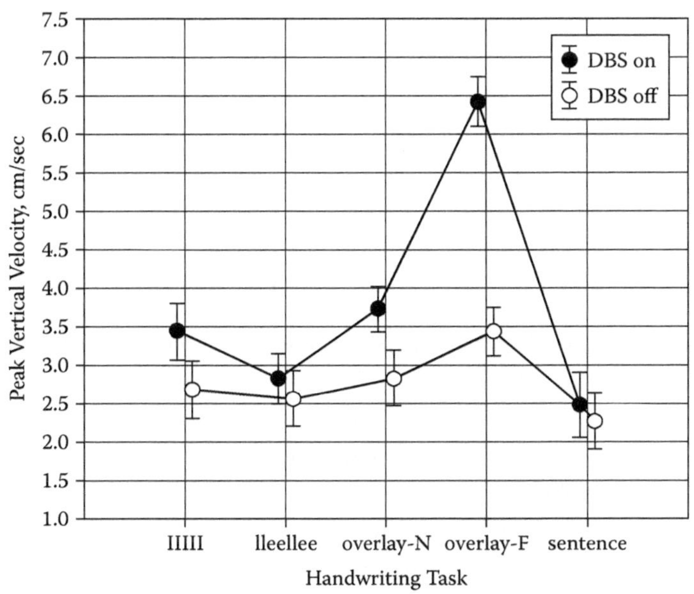

Figure 10.5 Means (with 95% confidence intervals) for peak stroke velocity for DBS on versus DBS off conditions across five handwriting tasks for a single PD subject.

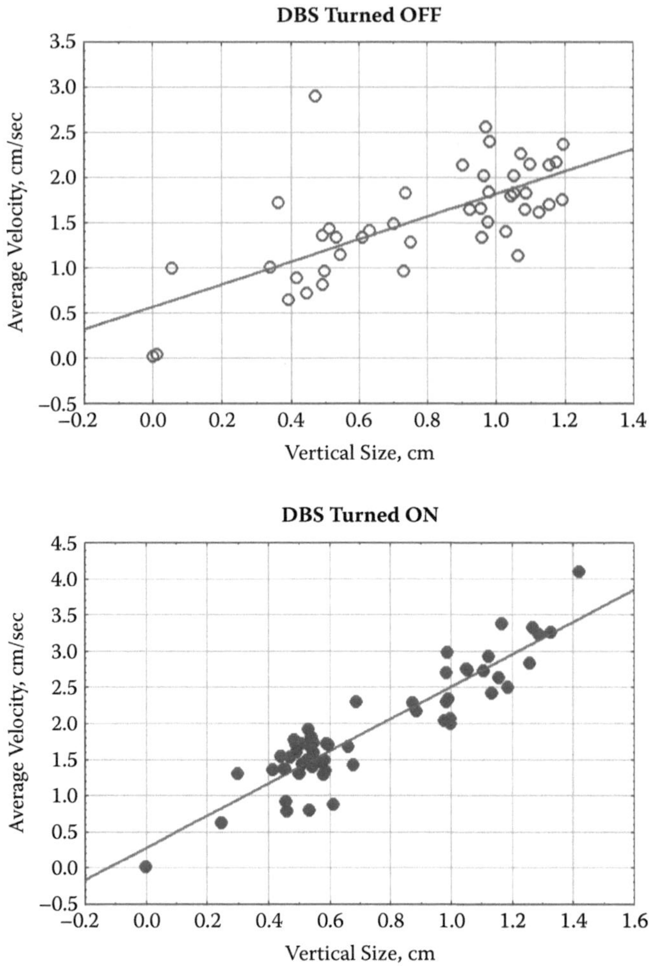

Figure 10.6 Results from the analysis of velocity scaling. Scatterplots show the relationship between stroke length and velocity for pen movements during cursive writing of alternating sequences of the letters *"llleellee"* with the DBS unit turned off (top) and on (bottom).

readily apparent that the length–velocity relationship is significantly stronger with fewer strokes falling outside the line of best fit. Thus, the patient demonstrated stronger adherence to the isochrony principle with the DBS unit turned on than when the unit was off.

To our knowledge, this is the first demonstration that systematic electrical stimulation of subcortical nuclei (i.e., the subthalamic nuclei) has a direct effect on the execution of a fundamental component of motor programming during human motor control. This demonstration using handwriting also confirms that the handwriting motor program may involve the same fundamental principles (isochrony) as other programmed movements. In summary, while the therapeutic effects of DBS on mobility, movement speed, and fine motor control are well recognized, this case presentation demonstrates that the benefits of DBS are also realized for handwriting. Mechanistically, these findings strengthen support for the importance of the basal ganglia, particularly the subthalamic nucleus, in handwriting motor control.

Essential Tremor

In his paper to the American Society of Questioned Document Examiners, Carney (1993) called attention to the distinction between ET and parkinsonian tremor as observed in handwriting. Unlike the resting tremor in PD, the tremor observed in ET patients is considered a postural tremor. That is, tremor amplitude increases during sustained posture or voluntary action such as handwriting. Handwriting is particularly vulnerable to the deleterious effects of ET because it involves both sustained posture (holding and gripping a writing instrument) and voluntary action.

Unlike the micrographic handwriting movements of PD, handwriting in ET is typically characterized by large amplitude strokes (Bhidayasiri 2005). Figure 10.7 shows the signatures from an individual prior to (a) and concurrent with (b) the development of ET. The contrast in vertical stroke length between the two samples is readily apparent. The contemporary signature (b) was obtained by one of the authors (MC), who observed that the frequency of up and down pen strokes was synchronized to the frequency of the tremor (approximately 7 Hz) such that the writer was unable to write his signature more slowly or faster than dictated by the tremor frequency. This phenomenon dramatically restricts the variability in pen stroke amplitude and timing normally available to the healthy writer.

As noted throughout this chapter, tremor occurs in many neurological conditions, including PD, ET, cerebellar disease, and certain forms of dystonia. Inevitably, the tremor progresses to the point where normal everyday functioning such as holding a fluid-filled cup, shaving, buttoning a blouse, cooking, and writing become difficult if not impossible. Most of these patients

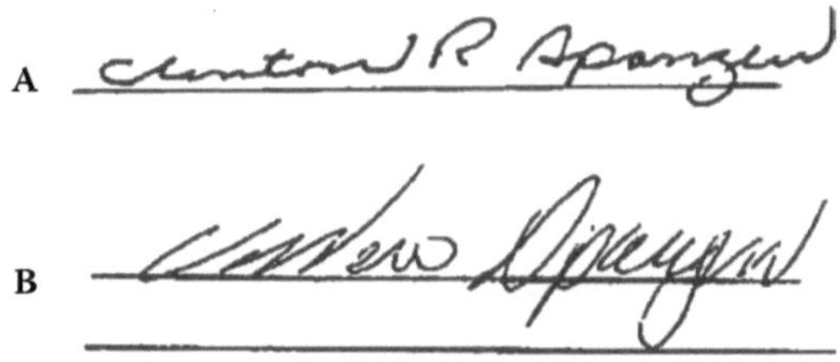

Figure 10.7 Signature samples from a single writer obtained prior to (A) and 15 years following (B) the diagnosis of essential tremor.

present multiple neurological signs throughout the musculoskeletal system. However, a subset of ET patients experience tremor only when writing—a condition known as primary writing tremor, or PWT (Rothwell, Traub, and Marsden 1979; Bain et al. 1995; Byrnes et al. 2005).

Rothwell and colleagues (1979) described PWT as a specific action (or kinetic) tremor characterized by an alternating pronation/supination tremor during writing that is not seen during other activity involving the forearm. PWT has two forms. Form A (task induced) is used to describe this tremor when it occurs only during handwriting; form B (positional) refers to this tremor during handwriting or when the person adopts the hand position normally used for handwriting. Controversy surrounds the classification of PWT; some argue that it is a subtype of ET with shared pathophysiology (Kachi et al. 1985; Jimenez-Jimenez et al. 1998; Modugno et al. 2002;), while others argue that it is a form of writer's cramp (Soland et al. 1996) or an independent tremor entity (Hai et al. 2010). For example, while ET is considered a progressive disease, PWT remains relatively stable over time, suggesting that ET and PWT may stem from different causal mechanisms (Bain et al. 1995). On the other hand, there is compelling evidence that both ET and primary writing tremor can be inherited[1] as an autosomal dominant trait, suggesting a shared pathophysiology (Bain et al. 1995). One important feature of distinguishing patients with PWT from the general ET population is the absence of resting tremor in PWT (Kachi et al. 1985; Bain et al. 1995).

When assessing the effect of ET on handwriting, clinicians routinely ask their patients to draw an Archimedes spiral. In a systematic study of spirographic drawing by ET patients in which arm posture and instructional conditions were varied, Ondo et al. (2005) found that tremor was rated as more severe when the writing arm was unsupported and when the patient was instructed to draw spirals between or on lines compared to supported posture and freehand spirals. This has implications for the evaluation of handwriting and signature variability. When variability among multiple signature samples is examined, it is important to consider whether the signature was written with or without spatial constraints. Specifically, writers with ET will likely exhibit a more deviant handwriting pattern (e.g., larger vertical amplitudes) when attempting to write within a confined space such as on a ruled line than when writing without such constraints.

Quantitative methods have been employed by researchers attempting to improve the reliability and sensitivity of tremor assessment. The two more common approaches include use of digitizing tablets to evaluate writing tremor and accelerometry to evaluate resting and postural tremor. Figure 10.8 shows raw samples of cursive *"e"*s and corresponding velocity curves associated with a single letter from a healthy writer (A and B) and a writer with tremor (C and D) from van Gemmert et al. (2003).

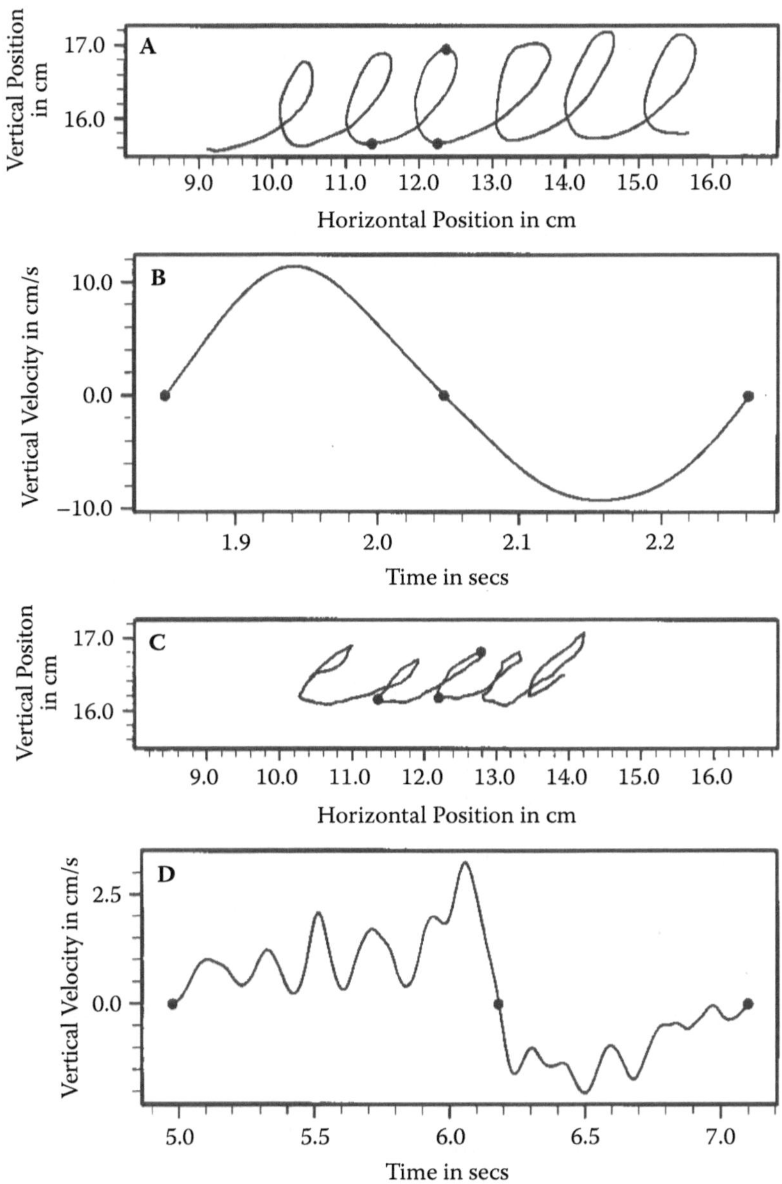

Figure 10.8 Raw samples of cursive "*e*"s and corresponding velocity curves associated with a single letter from a healthy writer (A and B) and a writer with tremor (C and D). From trace D, the tremor frequency is estimated to be 6 Hz, the typical tremor frequency of PD. ET tremor could be slightly higher. (From van Gemmert, A. et al. 2003. *Journal of Neurology, Neurosurgery, & Psychiatry* 74:1502–1508. With permission.)

A Kinematic Study of PD and ET Forged Signatures

An ongoing challenge within the FDE community is the authentication of signatures written by individuals with progressive neurological disease affecting fine motor skill. The FDE compares and assesses features (such as stroke length and slant, letter formation, connecting strokes, pen lifts, line quality, pen pressure, base alignment, hesitation, patching, retouching, and retracing) between the questioned and known signatures and then makes a subjective judgment as to whether the signature is genuine or not. This effort can be especially challenging when the questioned documents include samples written by an individual with a progressive movement disorder such as PD or ET. In cases where the writer exhibits parkinsonian micrographia or tremor, the static signature may be compromised, rendering evaluation ambiguous at best. More importantly, signatures of individuals with progressive disease change over time depending on the stage of illness and effectiveness of pharmacotherapy.

Two problems face FDEs tasked with evaluating signatures from individuals with PD or ET: validating their judgments, as required under the *Daubert* standard, and understanding judgment error as a means of improving reliability. The statistical approach to validating any questioned behavior requires classification of responses against an independent "gold standard." Unfortunately, in forensic document examination, there is no gold standard.

Recognizing this limitation, we undertook a small exercise to identify relevant kinematic elements that distinguish authentic from forged signatures written by individuals with micrographia or tremor. We view this as a first step in creating an independent gold standard of the features associated with tremulous or micrographic signatures.

Four patients volunteered to provide authentic signatures for this exercise. Two exhibited resting tremor (PD), one postural tremor (ET), and one had micrographia (PD). Each patient wrote his or her signature 10 times on a Wacom digitizing tablet using an inking pen. Following the collection of genuine signatures, an individual (generally a family member accompanying the patient) was asked to practice simulating the original signature and when ready, to forge the signature using the same apparatus. Kinematic measures of pen strokes were obtained using MovAlyzeR software. Sample genuine and questioned (forged) signatures are shown in Figure 10.9 for the three patients with tremor and one with parkinsonian micrographia.

Figure 10.10 shows the kinematic results comparing genuine with simulated signatures for each of the four patients. Forged signatures for the tremor samples were characterized by lower vertical stroke length (size) and higher vertical stroke amplitudes than those of genuine signatures.

Simulated signatures followed a similar pattern for loop surface (i.e., the total area derived from letters having open loops); however, the differences

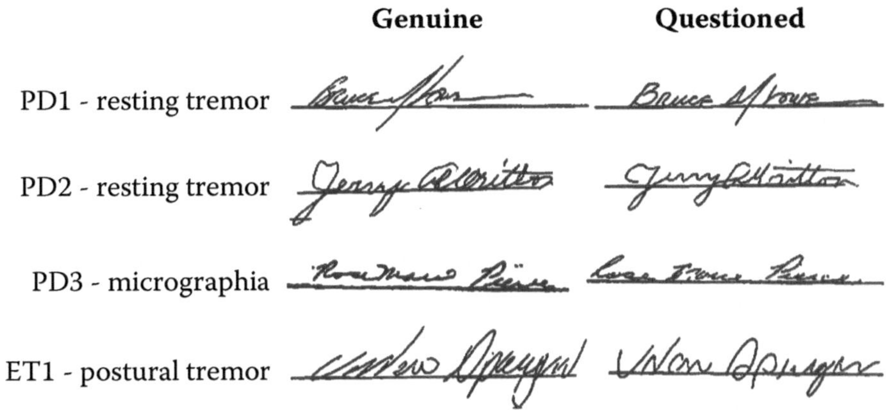

Figure 10.9 Genuine signatures from patients with PD or ET (left) and matching questioned signatures (right). Questioned signatures were subjected to analyses to identify kinematic strategies utilized by writers to forge signatures exhibiting micrographia and tremor.

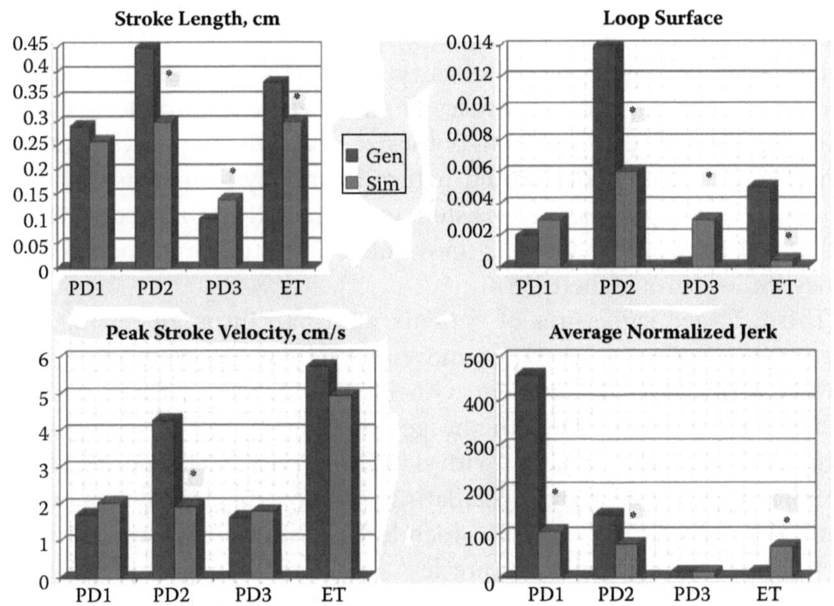

Figure 10.10 Kinematic results comparing genuine with simulated signatures for each of the four patients.

between genuine and simulated signatures for this parameter were striking for three of the four signatures. For example, the simulated signature of the ET patient had extremely low surface volume compared to the genuine signature; the opposite was true for the patient with micrographia. With the exception of one pair of signatures (resting tremor-PD2), simulated signatures were produced with similar stroke velocities as the original genuine signatures. Lastly, our results indicate that attempts to simulate signatures of patients with parkinsonian resting tremor were written with significantly smoother pen movements than the original signatures were. Interestingly, pen movements associated with the genuine signature from our ET patient were relatively smooth, whereas the simulated traces were characterized by greater dysfluency.

A few patterns emerged from this demonstration that may shed light on strategies used by forgers when simulating pathological signatures. First, when attempting to simulate a micrographic signature, the forger will likely have difficulty executing handwriting movements as small as those of the genuine signature. This may be due to inflexibility of the handwriting motor program to reduce the size of complex finger and wrist movements voluntarily.

Second, when attempting to simulate tremor, regardless of the type of tremor, forgers tended to decrease the size and surface area of the signatures. One could speculate that this is an adaptive mechanism necessary to mimic voluntarily the involuntary oscillating pen movements that characterize tremor. Since pathological tremor is typically characterized by movement oscillations with higher frequencies than movements produced naturally during handwriting, we would hypothesize that reducing movement extent is an effective strategy one would naturally employ to increase movement frequency. This hypothesis is consistent with the well established inverse biomechanical relationship between movement amplitude and movement frequency (Stiles 1976; Hallett 1998).

Third, forged signatures of patients with parkinsonian resting tremor were written with smoother pen movements than the genuine signatures. Conversely, forgeries of a patient with ET tremor were written with more dysfluent pen movements than the genuine signature. This suggests that attempts to simulate a signature with significant postural tremor result in the writer having to model rapidly oscillating pen movements that are not part of a natural handwriting program, which leads to inefficiency and dysfluency. The finding that simulations were less dysfluent than genuine signatures from patients with resting tremor may have little to do with their tremor and more to do with other motor aspects of parkinsonism, such as bradykinesia and rigidity. Handwriting in PD is likely to be dysfluent for any number of reasons—such as hesitations caused by rigidity, delayed initiation times, or tremor—that are difficult for a healthy writer to simulate.

Caution should be exercised when interpreting these preliminary findings. These simulations and subsequent kinematic observations were derived from only four pairs of writers with limited generalizability. Nonetheless, our observations are preliminary and serve to guide future research to elucidate the motor control strategies employed by writers attempting to forge pathological signatures.

Progressive Supranuclear Palsy

As noted in Chapter 4, the motor, cognitive, and behavioral features of progressive supranuclear palsy (PSP) overlap with those of PD (Cordato et al. 2006). While fewer PSP than PD patients exhibit micrographia, approximately 25% of the PSP patients have impaired handwriting in some form.

Recently, one of the authors (MC) was asked to evaluate the handwriting of an 80-year-old individual with PSP for the purpose of shedding light on the complex time course of this devastating disease.[2] Here, we describe a neuropathologically confirmed case of atypical progressive supranuclear palsy presenting with corticobasal syndrome. Handwritten signatures were obtained at regular intervals from the patient for 3 years prior to death to document change in motor function. This case sheds light on the differential effects and time course of progressive subcortical and cortical disease on handwriting in a single individual.

At age 80 the patient presented with weakness, gait disturbance, an inability to write, difficulty with fine motor control, and difficulty holding objects with the right hand. Her gait was shuffling in appearance with a stooped stance and restricted swing of the right arm. Initially, the patient had a mild right-hand action tremor, which developed later in the left hand. She was treated with Sinemet 25/100 CR TID off and on throughout the last 3 years of her life. She reported no prior exposure to neuroleptics or other medications known to affect handwriting. Follow-up physical exam revealed severe impairment of extraocular movements, especially with vertical gaze. Except for the intrinsic muscles of the right hand, strength was normal.

On exam, the patient was found to have uncontrolled movements of the right hand, increased tone throughout, and a stooped, shuffling gait. The presence of postural instability, shuffling gait, and supranuclear gaze palsy was consistent with a diagnosis of Richardson syndrome. As the disease progressed, she had difficulty performing most commands involving the right hand suggestive of dyspraxia. The clinical picture near the end of her life was consistent with corticobasal syndrome (CBS).

The referring forensic document examiner was keenly aware of the changes in the patient's handwriting and obtained handwritten signatures at relatively equal intervals over the last several years of the patient's life. Figure 10.11 shows samples of these signatures from 2004 (1 year after diagnosis) through 2009 (1 year prior to death). Signature A was obtained prior to the presentation of any clinical signs suggesting a progressive neurological disease. Signature B was obtained at a time when the patient exhibited her first signs of weakness and gait disturbance. This signature remains legible;

	Date	Sinemet	Signature
A	12-03-2004	Off	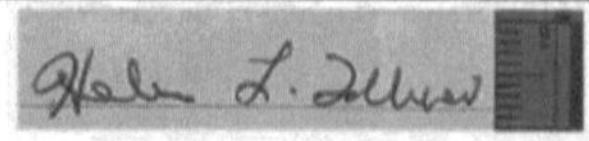
B	11-30-2007	On	
C	08-12-2008	On	
D	12-13-2008	Off	
E	07-02-2009	On	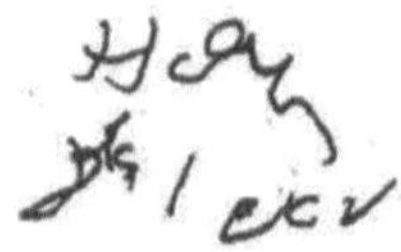
F	07-24-2009	On	

Figure 10.11 Handwriting samples from a patient initially diagnosed with progressive supranuclear palsy that later progressed into corticobasal syndrome.

however, it no longer adheres to an imaginary baseline plane as healthy signatures do. Signatures C and D were obtained 1 year after the initial diagnosis. These signatures show clear micrographia, a classic sign in PD, and also demonstrate the effects of dopamine replacement therapy on handwriting.

While both signatures appear micrographic, signature D shows more severe micrographia at a time when the patient was off Sinemet. Signature D is largely illegible; however, the sequencing and spacing are preserved. As with the 2007 sample (B), the signature fails to follow an imaginary writing plane. Signatures E and F were obtained approximately 1 year later. Interestingly, these signatures are no longer micrographic, but essentially illegible. Character sequencing and spacing in both vertical and horizontal axes are disrupted. The signatures are disorganized with first and last names appearing on different horizontal planes. For some characters (e.g., signature E, first few characters of the last name), the trace is more of a scribble with no resemblance to an identifiable letter.

Qualitative analysis of the patient's handwriting revealed changes over a 2-year period that paralleled the progression of the neurodegeneration. Micrographia early in the course of the disease progression was consistent with an initial diagnosis of PD. Furthermore, the beneficial effects of dopamine replacement therapy on handwriting, as shown in signatures C and D of Figure 10.11, were consistent with PD. Later signatures exhibited character sequencing and spatial disorganization typically seen in apraxic handwriting associated with CBS (Murray et al. 2007).

The signatures obtained over a 2-year period from the patient reflect an atypical progression of PSP. Once the disease began to impact handwriting, the signatures were micrographic and resembled parkinsonism. Signatures obtained from the patient on and off dopamine replacement therapy are consistent with a distinct parkinsonian phenotype in PSP. Within the last year of the disease, the patient's signatures became less micrographic and more disorganized and dyspractic as the disease began to involve higher cortical brain centers. The late signatures were disorganized spatially and included characters that had little or no resemblance to English letters.

This case demonstrates that careful documentation of the variability in handwriting impairment over time can provide a unique opportunity to observe the differential effects of progressive neurological syndromes, particularly in cases involving subcortical and cortical motor degeneration.

Huntington's Disease

An overview of the pathology underlying Huntington's disease (HD) was presented in Chapter 4. Recall that HD is an inherited, progressively disabling disorder that causes problems with behavioral control, cognition, and motor function. The movement disorder in HD is a hyperkinetic disorder characterized by dyskinesia, the random involuntary movements of the limbs and trunk. Handwriting movements in HD are characterized by the interruptions and excessive number of velocity and acceleration reversals within stroke and excessive variability between strokes. Patients with advanced forms of HD may exhibit signs of parkinsonism, including tremor and micrographia, as the disease progresses to degeneration of nigrostriatal dopamine projections.

There have been sparingly few published papers on handwriting characteristics in HD. Work by Phillips and colleagues (1994, 1995; Slavin et al. 1999) constitute the bulk of this literature. Phillips et al. (1994) employed kinematic analyses to characterize handwriting movements in 12 patients with HD. Subjects were instructed to write the letter "*l*" at a specified size using linked cursive script. Samples were digitized and nearly a dozen kinematic parameters were automatically extracted from the digitized samples. Of particular relevance to this chapter was the stroke-to-stroke consistency and smoothness of pen movement. The investigators reported that handwriting movements in HD were characterized by longer stroke durations; greater variability in stroke amplitude, duration, and velocity; and an increase in the number of submovements (an index of handwriting smoothness and efficiency) compared to comparably aged, healthy subjects. The investigators did not find any associations between the handwriting kinematic abnormalities and clinical status or medication, suggesting weak predictive value of handwriting kinematics as markers of disease progress in HD.

In a subsequent study, Phillips et al. (1995) examined whether these impairments worsened during continuous sequential handwriting as is observed in PD micrographia. Given that the basal ganglia pathophysiology in HD overlaps with PD, at least in advanced cases, one could hypothesize that handwriting in HD should contain some micrographic features. While the investigators observed that stroke duration increased progressively with continuous writing, there was no evidence of progressive decrease in stroke length as there is in PD micrographia. Nonetheless, at least 1 of the 12 HD patients studied by Phillips et al. (1994) exhibited clear micrographia. A few years later, Iwasaki et al. (1999) reported micrographia in a 72-year old HD patient, lending support to the notion that HD and PD share a common basal ganglia pathology that could manifest as micrographia.

To clarify further whether abnormal handwriting kinematics in HD may be an early sign signaling the onset of HD clinical symptoms and therefore useful in the clinical management of HD patients early in the course of the disease, we conducted a study of unaffected family members of HD patients. It is well known that HD is caused by an autosomal dominant mutation on either of an individual's two copies of a gene called Huntingtin (HTT); this means that any child of an affected parent has a 50% risk of inheriting the disease.

All humans have the Huntingtin gene, which codes for the protein Huntingtin. Part of this gene is a repeated section called a trinucleotide repeat (CAG repeat number), which varies in length between individuals and may change length between generations (Walker 2007). When the length of this repeated section reaches a certain threshold, it produces an altered form of the protein, called mutant Huntingtin protein, which increases the decay rate of certain types of neurons. The number of repeats is related to how much this process is affected and the age when symptoms may appear. For example, 36–40 repeats are associated with a late onset form with slower progression of symptoms, whereas with very large repeat counts, HD symptoms occur at a very young age and progress rapidly (Nance and Myers 2001).

In our study, handwriting movements from 18 asymptomatic family members of HD patients, three symptomatic HD patients, and 10 healthy comparison subjects were examined. The 18 family members were genetically tested to evaluate the trinucleotide repeat on chromosome 4 at 4p16.3 (where the HTT gene is located). Of the 18 subjects, 6 showed abnormal repeats (at

Table 10.3 Results from Kinematic Analyses of Pen Stroke Movements during Continuous Production of the Letter "*l*" Written within a 4 cm Vertical Boundary for Two Groups of Subjects at Risk for HD[a], Symptomatic HD Subjects[b], and Normal Comparison Subjects[c]

	Stroke Duration (ms)	Stroke Length (cm)	Stroke Velocity (cm/s)	Normalized Jerk
AR-Pos	247 (91)	2.00 (0.22)	10.74 (4.59)	18.67 (8.40)
AR-Neg	317 (116)	2.03 (0.37)	8.76 (3.11)	27.78 (21.59)
HD	435 (213)	1.30 (0.19)	4.69 (1.95)	120.18 (38.60)
NC	276 (105)	1.93 (0.34)	9.45 (4.09)	20.41 (12.00)

Note: Shown are the mean scores (with standard deviations).
[a] AR-Pos ($n = 6$) and AR-Neg ($n = 12$).
[b] $n = 3$.
[c] $n = 10$.

high risk for HD) while 12 did not. Subjects were asked to write the letter "*l*" continuously in cursive script within 1, 2, and 4 cm vertical boundaries. Samples were written on a Wacom digitizing tablet using an inkless pen. The task was repeated five times. Digitized pen movements were processed using MovAlyzeR software from which several kinematic parameters were extracted, including vertical stroke duration, amplitude, velocity, and normalized jerk (a measure of smoothness). The examiner was blinded to the genetic results at the time of the handwriting task.

Results for the 2 cm task condition are shown in Table 10.3. Results failed to demonstrate statistically significant differences on any handwriting kinematic parameter between family members who tested positive versus those who tested negative for the HD genetic mutation. However, symptomatic HD subjects exhibited significantly greater normalized jerk than any of the other subject groups for the 1 cm ($F = 11.85$; $p < 0.0001$), 2 cm ($F = 21.82$; $p < 0.0001$), and 4 cm ($F = 16.88$; $p < 0.0001$) tasks. Symptomatic HD subjects exhibited significantly lower vertical stroke heights than any of the other subject groups ($F = 4.10$; $p < 0.05$).

Our findings indicated that, at least for the subjects of this study, kinematic analyses of cursive letter writing were not useful as early makers of symptom onset in subjects testing positive for the HD gene. Nonetheless, our results from kinematic analyses of handwriting in symptomatic HD subjects confirm previous observations (Phillips et al. 1995; Iwasaki et al. 1999) that handwriting in HD is highly dysfluent and that symptomatic HD subjects may exhibit micrographia.

Alzheimer's Disease

The principal handwriting impairment in early Alzheimer's disease (AD) has a cognitive-linguistic basis composed of lexical or semantic errors, word selection, and phonological substitutions (Rapcsak et al. 1989; Platel et al. 1993). Handwriting among patients with mild or early AD generally shows no lexical or graphic motor impairment (Croisile et al. 1995; Hughes et al. 1997). However, later in the course of the disease, writing samples show more graphic motor disturbances (Hughes et al. 1997). Studies of the lexical-semantic aspects of handwriting impairment in AD conclude that the pattern of impairment is similar to that observed in focal brain damage such as aphasia (Luzatti et al. 1998; Luzatti, Laiacona, and Hagáis 2003).

As noted in Chapter 4, handwriting movements of most AD patients remain relatively preserved throughout their lives. While there have been numerous published works characterizing the linguistic aspects of handwriting impairment in AD (e.g., Rapcsak et al. 1989; Hughes et al. 1997; Luzzatti et al. 2003; Silveri, Corda, and Di Nardo 2007), few studies have focused on the motor aspects of handwriting disturbance specific to AD (Slavin et al. 1995, 1999; Schröter et al. 2003; Werner et al. 2006; Yan et al. 2008).

Slavin et al. (1999) studied handwriting efficiency in AD by examining the consistency of handwriting movements. Sixteen AD patients (ranging in severity from mild to severe dementia) were asked to write four cursive lower-case letter "*l*"s on a digitizing tablet using an inkless pen. Consistency, defined as a signal-to-noise ratio or the mean stroke parameter divided by its standard deviation, was calculated for pen stroke duration, amplitude, and velocity. The task was repeated under normal and reduced visual feedback. The investigators found that while stroke duration and amplitude were relatively intact, AD patients exhibited less consistent movements than healthy comparison subjects. Medication was ruled out as a contributing factor on the basis that unmedicated patients exhibited similar levels of stroke variability. Performance was more variable under reduced visual feedback, suggesting a primary motor programming disturbance (assuming closed loop control for handwriting).

On the basis of previous work in PD and HD (Phillips, Stelmach, and Teasdale 1991; Phillips et al. 1994), Slavin and colleagues concluded that the handwriting movement patterns in AD were similar to those observed in HD reflecting similar basal ganglia pathophysiology. It is interesting to note that unlike PD patients, HD and AD patients exhibit disproportionate cognitive impairment relative to their motor involvement. Thus, overlapping handwriting patterns in AD and HD could very well stem from a disturbance in the cognitive aspects of the handwriting motor program.

Schröter and colleagues (2003) evaluated handwriting movements in patients with AD, mild cognitive impairment (MCI), and healthy subjects to test whether these groups differed systematically on measures of handwriting kinematics and whether handwriting dysfunction can be used to differentiate patients with mild forms of cognitive impairment from those with AD. Subjects were instructed to draw concentric superimposed circles as fast and fluently as possible with and without a distraction task. Measures of handwriting speed (frequency or number of circles/second, velocity, and variability in velocity between strokes) and smoothness (changes in velocity direction) were extracted. The investigators found that AD patients exhibited significantly greater variability in velocity than MCI and healthy subjects; however, no differences were found in movement speed or frequency. Dementia severity was not correlated with handwriting kinematics in AD, suggesting more of a pure motor programming deficit. These findings are consistent with Slavin et al. (1999) and strengthen the hypothesis that handwriting movements in AD are characterized by an increase in stroke-to-stroke variability.

A recent study by Yan et al. (2008) also focused on kinematic differences in handwriting movements between AD and MCI patients. Unlike previous research that employed letter writing tasks, subjects in the Yan et al. study were instructed to move a stylus quickly between two dots using either a two-stroke (two back and forth progressions) or four-stroke (four back and forth progressions) handwriting movement. Measures of movement time and movement jerk (a measure of smoothness) were obtained for each stroke. They reported that both patient groups exhibited slower and less smooth movements than healthy comparison subjects. As with prior studies (Slavin et al. 1999 and Schröter et al. 2003), movement times were more variable and handwriting movements less smooth than those of healthy controls. Yan et al. observed both impaired handwriting movement duration and increased movement dysfluency in their MCI patients. These results underscore the difficulty in distinguishing AD from MCI on the basis of handwriting kinematic analyses alone.

While there is a clear need for more research, handwriting in AD may be characterized by the preservation of kinematic features such as speed, stroke duration, and size (adjusted for age) with increased variability and loss of smoothness and fine control. Debate remains as to whether the decline in handwriting in AD reflects the pathological change in fronto-cortical integrity, giving rise to cognitive and psychomotor deficits (Slavin et al. 1999), or pathological change in subcortical basal ganglia integrity, giving rise to parkinsonian features as in dementia with Lewy bodies (DLB). It is likely that both processes are involved.

To address this question, we conducted a pilot study of handwriting kinematics in AD patients with and without probable DLB.[3] Our goals were to determine whether kinematic analyses of handwriting movements support previous literature that handwriting movements are preserved in AD and to identify kinematic parameters that might distinguish AD from DLB. We employed our standard laboratory assessment of handwriting (as described throughout this book). Briefly, subjects were instructed to draw concentric circles, write series of the letter "*l*" and alternating "*lleelle*," and

Table 10.4 Means (Standard Deviation) for Selected Handwriting Kinematic Variables for Healthy Subjects and AD Patients for Sentence Writing

	Healthy writers (n = 7)	AD writers (n = 9)
Stroke duration, ms	279 (36)	395 (202)
Vertical stroke length, cm	1.04 (0.18)	1.09 (0.89)
Vertical stroke velocity, cm/s	7.56 (1.58)	6.44 (1.78)
Number of acceleration peaks/stroke	2.99 (0.44)	4.21 (2.75)

write the sentence "Today is a nice day" using an inkless pen on a Wacom digitizer. MovAlyzeR software was used to acquire and process the kinematic variables for each pen stroke.

Preliminary results were available from nine AD patients (mean age of 74.8 years) and seven healthy control subjects of comparable age and gender (mean age of 71.8 years). Table 10.4 shows the results for the two subject groups from the sentence-writing task. While AD writers exhibited longer stroke durations, lower stroke velocities, and greater number of acceleration peaks (inversions) per stroke, these means were not significantly different from those of healthy writers. Similar results were obtained from the repetitive circle and sequential letter writing tasks.

The results show that AD patients were more variable as a group than healthy writers, suggesting that some AD patients may have impaired handwriting. One likely source of this variation could be the presence of motor signs consistent with the provisional diagnosis of DLB. Motor status was documented in five of the nine AD patients using the UPDRS. Three of the five AD patients exhibited motor impairment suggestive of parkinsonism and were therefore given the provisional diagnosis of DLB. A comparison of the kinematic scores for the DLB versus the two known non-DLB patients is shown in Figure 10.12.

These results from a very small sample support the notion that dementia patients with probable DLB based on clinical criteria are likely to exhibit handwriting impairments that resemble PD. Specifically, significantly slower movement velocities characterized handwriting in DLB. The longer stroke durations, decreased stroke length, and increased number of acceleration inversions did not reach statistical significance due to the small sample size and low statistical power to detect group effects. Nonetheless, the AD patients in general exhibited a 40% increase in the number of acceleration inversions per stroke compared with healthy writers (see Table 10.4); DLB+ patients exhibited a 75% increase compared with DLB– patients. These findings indicate that handwriting may be impaired in AD patients—particularly those who meet clinical criteria for DLB—and that the nature of this impairment may not have a solely cognitive/linguistic basis.

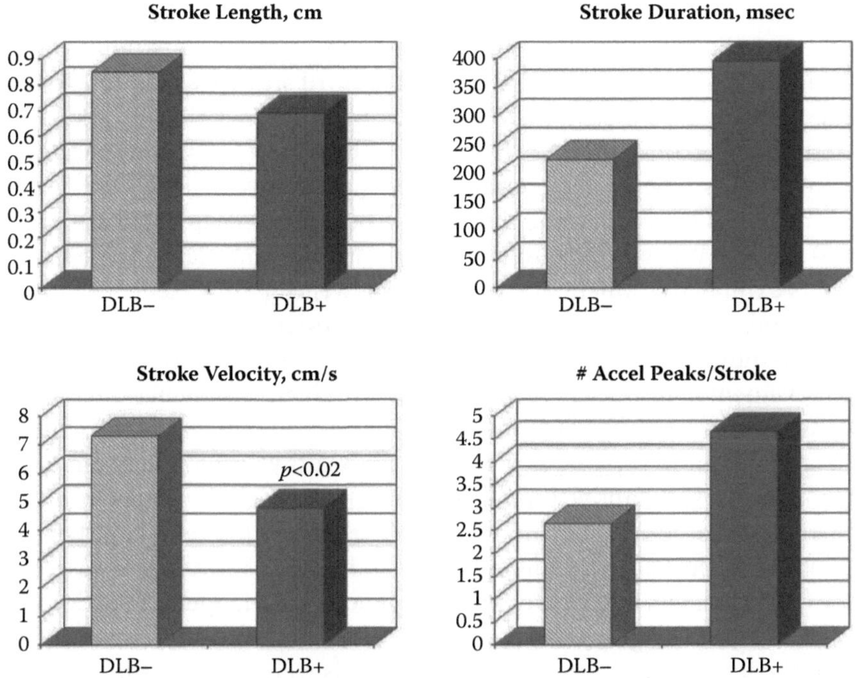

Figure 10.12 Comparison of handwriting kinematic variables from sentence writing by five AD patients grouped according to those who met clinical criteria for dementia with Lewy bodies (DLB+; $n = 3$) and those who did not (DLB–; $n = 2$).

Summary

This chapter reviewed empirical research from our laboratory and others on the effects of neurological disease on handwriting kinematics. Research on handwriting movements among individuals with neurological disease can inform underlying pathological mechanisms responsible for the disease and can provide a record of change in disease progress or benefits of treatment. We summarized findings from studies of patients with Parkinson's disease (PD), essential tremor (ET), Huntington's disease (HD), progressive supranuclear palsy (PSP), and Alzheimer's disease (AD) with and without DLB. While time course and clinical management differ for these conditions, there is overlap in their neurochemistry and pathophysiology, particularly with regard to subcortical brain regions that govern motor control. Given the overlapping brain regions thought to be involved in the expression of motor problems of these conditions, it is reasonable to speculate that they would also show overlapping patterns of abnormal handwriting kinematics.

Three prospective studies from our laboratory were presented in this chapter. The aim of two of these studies was to examine whether the known pathological processes underlying movement disorders in two common conditions (Parkinson's disease and Alzheimer's disease) extend to handwriting movements as well. The aim of the third study was to identify kinematic features that distinguish genuine signatures written by individuals with tremor and micrographia from attempts to forge these same signatures.

In the PD study, we used sophisticated quantitative methods of handwriting kinematics to confirm the therapeutic benefits of deep brain stimulation (DBS) for PD. We were able to show that with the stimulator turned on, PD patients exhibited dramatic increase in handwriting movement speed, stroke length, and smoothness. In the AD study, we tested whether kinematic analyses of handwriting movements support previous literature that handwriting movements are preserved in AD and to identify kinematic parameters that might distinguish AD from DLB. The results indicated that while AD patients were more variable as a group than healthy writers, they did not differ on measures of handwriting kinematics from healthy writers as a group. Nonetheless, some AD patients may have impaired handwriting. Dementia patients with probable DLB exhibited handwriting movements resembling those of PD patients (i.e., slower movement velocities, longer stroke durations, decreased stroke length, and increased number of acceleration inversions).

In the forgery study, untrained individuals were asked to forge these signatures. By comparing the kinematic features of the genuine with those of the forged signatures, patterns emerged from this demonstration that may shed light on strategies used by forgers when simulating pathological signatures. First, when attempting to simulate a micrographic signature, the forger had difficulty executing handwriting movements as small as the genuine signature. This may be due to inflexibility of the handwriting motor program to reduce the size of complex finger and wrist movements voluntarily.

Second, when attempting to simulate tremor, regardless of the type of tremor, forgers tended to decrease the size and surface area of the signatures. Since pathological tremor is typically characterized by movement oscillations having higher frequencies than movements produced naturally during handwriting, we would hypothesize that reducing movement extent is an effective strategy one would naturally employ to increase movement frequency.

Third, simulations of genuine signatures from patients with parkinsonian resting tremor were written with smoother pen movements than the genuine signatures, suggesting that attempts to simulate a signature having significant postural tremor causes the forger to produce rapidly oscillating pen movements that are not part of a natural handwriting program; this leads to inefficiency and dysfluency.

While these studies help elucidate the impact of neurological disease on handwriting, distinguishing the effects of healthy aging from neuropathological conditions remains an ongoing challenge for both neuroscientist and document examiner.

Notes

1. Support for the genetic susceptibility of ET comes from an interesting historical review by Louis and Kavanaugh (2005) on the tremor of John Adams, the second president of the United States. In their thorough review of historical documents and personal letters, the authors compile a fascinating timeline showing signs of a low-amplitude kinetic tremor beginning when John Adams was 25 years of age, which progressed through life. Supportive documents revealed that his cousin (Samuel Adams) as well as his son (John Quincy Adams) also had tremor (Louis 2001; Paulson 2004) consistent with the diagnosis of ET.
2. The authors acknowledge the referral by Diane Tolliver, a senior forensic document examiner from Indianapolis, Indiana, and her thorough documentation and appraisal of the patient's signatures.
3. We acknowledge the support of the UCSD Alzheimer Disease Research Center for assistance in subject recruitment and collection of the handwriting data. Support for this research was provided by a grant from The National Institute of Aging (AG05131).

11 Effects of Psychotropic Medications on Handwriting

Introduction

One significant challenge that clinicians face in managing patients with mental illness such as psychosis, dementia, or severe depression is how to balance the therapeutic effects against the countertherapeutic effects of powerful psychotropic medications. Clinicians and researchers have searched for means to detect subtle changes in the neuromotor system attributable to these medicines to monitor the emergence of side effects in patients treated with psychotropic medications.

Interestingly, in the late 1950s and early 1960s handwriting was considered an ideal candidate for such a monitoring system. Haase (1961) was the first to demonstrate a relationship between clinical effectiveness of neuroleptic medications for treating psychosis and their side effects, using handwriting analysis. Haase noted that as neuroleptic dosage increased, patients showed parkinsonism. Handwriting for these patients became slowed (bradykinesia) and decreased in size (micrographia) as neuroleptic dose was increased. When the dosage was decreased, the handwriting disturbances disappeared, as did the therapeutic effects of the medication. This relationship was referred to as the "neuroleptic threshold," defined as the minimum dose a patient needs to obtain clinical efficacy while minimizing any of these sedating side effects.

Since then, clinicians have considered the extrapyramidal motor system as a reliable window into neuroleptic actions on the mesolimbic emotional system. Figure 11.1 is from a series of examples published by Haase and Janssen (1965) demonstrating the sensitivity of handwriting analysis as an objective measure of identifying an optimal dose of a neuroleptic drug. As can be seen in these examples, the posttreatment handwriting samples show micrographia—an indication of the dopamine blocking properties of effective antipsychotics available in the 1960s.

The effects of prescription drugs and medication on handwriting are of considerable importance to the forensic document examiner (FDE). One of the earliest reports to appear in the forensic science literature was a paper by Legge et al. (1964). The investigators described the effects of nitrous oxide (a

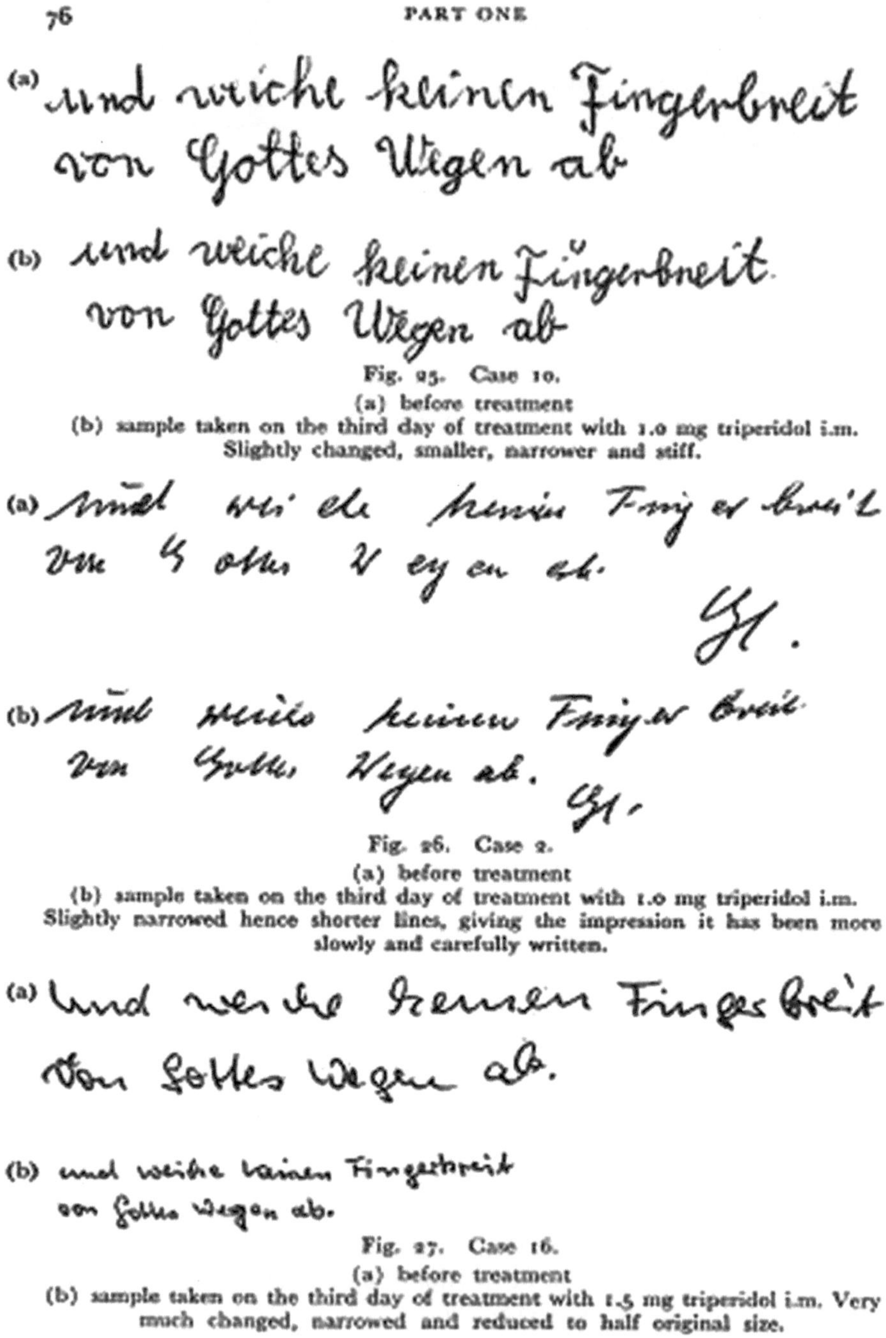

76 PART ONE

(a) und weiche keinen Fingerbreit von Gottes Wegen ab

(b) und weiche keinen Fingerbreit von Gottes Wegen ab

Fig. 25. Case 10.
(a) before treatment
(b) sample taken on the third day of treatment with 1.0 mg triperidol i.m. Slightly changed, smaller, narrower and stiff.

Fig. 26. Case 2.
(a) before treatment
(b) sample taken on the third day of treatment with 1.0 mg triperidol i.m. Slightly narrowed hence shorter lines, giving the impression it has been more slowly and carefully written.

(b) und weiche keinen Fingerbreit von Gottes Wegen ab.

Fig. 27. Case 16.
(a) before treatment
(b) sample taken on the third day of treatment with 1.5 mg triperidol i.m. Very much changed, narrowed and reduced to half original size.

Figure 11.1 Sample plate from the Haase and Janssen book on actions of neuroleptic drugs. (Haase, H. J., and Janssen, P. A. J. 1965. *The Action of Neuroleptic Drugs: A Psychiatric, Neurologic, and Pharmacologic Investigation.* Chicago: Year Book Medical Publishers. With permission.)

central depressant) on five measures of handwriting, including vertical height of lower-case letters, height of up and down strokes, peak vertical height, horizontal length, and spatial distribution. Fifty writers were randomly assigned to five dose groups. The investigators observed a significant dose-related increase in the vertical height of cursive script, particularly for lower-case letters. Increases in baseline (horizontal) length were the most consistent finding.

The investigators proposed two mechanisms to account for their findings. First, nitrous oxide may alter neuromuscular control by limiting the ability of the writer to produce small movements. An alternative mechanism involves altered perception leading to the distorted kinesthetic feedback. Nitrous oxide has three main clinical uses in humans, each with complementary mechanisms of action. As an anxiolytic, the effects of N2O are mediated by enhanced activity of inhibitory GABA receptors (see Chapter 1 for review of GABA pathways and motor control). Its analgesic effects are linked to the interaction between the endogenous opioid and the descending noradrenergic system, while the euphoric effects are mediated by stimulation of the mesolimbic dopamine pathways. Given these pharmacological actions of nitrous oxide, the latter explanation seems more plausible—particularly the GABAergic mechanism, which could decrease sensorimotor input to the descending thalamocortical motor pathway. These findings have implications for understanding the effects of other central depressants (such as neuroleptics and anxiolytics) on handwriting.

In this chapter we summarize some of the early work on neuroleptic effects on handwriting and how handwriting was used to optimize pharmacotherapy in patients with severe mental illness. Following this review, we turn our attention to the neurobiology of psychotropic-induced movement disorders in general and handwriting impairment in particular. We then describe recent empirical research from our laboratory and others on specific changes in handwriting kinematics associated with a variety of commonly used psychotropic medications.

Empirical Research on Effects of Psychotropics on Handwriting Kinematics

Following the extensive work by Haase and Janssen that began in the 1950s and extended through the 1970s (Haase 1978), there was very little research activity in the area of handwriting as a biomarker of antipsychotic toxicity. The field witnessed a rebirth following a paper by Gilmour and Bradford (1987), who described the handwriting of patients who were being treated for schizophrenia. While they reported that use of antipsychotic drugs by this population led to alterations in handwriting, the effects were highly variable across patients and medications. The investigators reported alterations in line quality (a measure of smoothness and continuity), size of handwriting, and letter formation. The investigators observed handwriting impairment in only 20% of their cases. However, no single drug group could account for the handwriting distributions across patients, suggesting variable individual response to these medications. For example, in some of their patients, handwriting alterations appeared early in the course of treatment and then remitted; in others, these disturbances persisted.

While the Gilmour and Bradford (1987) study was remarkable in its ability to record extensive psychiatric and medication histories from a large sample of psychiatric inpatients to address an important problem facing FDEs, the study was unable to identify consistent generalizable findings. This weakness is inherent in many studies based on subjective qualitative methods to examine complex handwriting movements.

More recent studies on the psychotropic effects on handwriting have utilized quantitative methods and as such are more likely to yield generalizable findings that could inform the FDE community. In the following paragraphs, we review studies employing methods for quantifying handwriting kinematics in patients treated with a wide range of medications including antipsychotics, antidepressants, anticholinergics, and anxiolytics.

The use of handwriting to assess antipsychotic-induced motor side effects has been the focus of research primarily in Europe (Haase 1978; Gerken et al. 1991; Kuenstler et al. 1999, 2000). Gerken et al. (1991) examined whether handwriting movement size (i.e., area encompassed by handwriting) could predict treatment response in their schizophrenic patients. The investigators reported that treatment with antipsychotics led to reduction in the overall size of the handwriting samples (defined as a 13% reduction in the overall size, or area, of 50% or more of the handwriting samples) in about one-third of the treatment responders. However, most of the treatment nonresponders also exhibited reduction in handwriting area, suggesting that handwriting may not be an effective predictor of treatment response. Rather, the authors concluded that handwriting parameters might be better suited for evaluating neurological side effects of neuroleptic medication than predicting treatment response using standard observer rating scales.

Kuenstler et al. (1999) used positron emission tomography to examine the relationship between reduction in handwriting size (expressed by area) and dopamine D2 receptor occupancy in schizophrenic patients before and after treatment with drugs (haloperidol, clozapine, or risperidone). Two important findings emerged from their work. First, they found reductions in handwriting size in all subjects following treatment, regardless of the medication type. A second finding was the highly significant linear relationship between D2 receptor occupancy and reduction in handwriting area. The authors concluded that analysis of handwriting size might be well suited for evaluating neurological side effects of neuroleptic medications. Findings from these and other published studies of handwriting demonstrated that antipsychotics impart observable changes in handwriting. Moreover, these changes were not limited to conventional antipsychotics.

We recently completed a large-scale, multisite study of handwriting kinematics in psychosis patients treated with a variety of psychotropic medications (Caligiuri et al. 2009, 2010). These studies were designed to examine whether a quantitative procedure for assessing handwriting movements could be used to distinguish among the newer, less toxic second-generation antipsychotics.

Our complete handwriting battery included 15 different writing tasks varying in vertical size and pattern complexity for both dominant and non-dominant hands and normal and faster writing speeds. The full battery of writing patterns included (1) cursive loops, (2) continuous circles, (3) a complex cursive loop sequence, and (4) a sentence: "Today is a nice day." All tasks were repeated three times each at 1, 2, and 4 cm vertical stroke heights except the sentence and the high-speed circles, which were produced only at the 2 cm vertical stroke size. The subjects performed all replications of one task before moving to the next task. The sequence of tasks was random. The duration of the handwriting test was about 20 minutes. Table 11.1 summarizes the subject characteristics of this study. We include only those results from patients treated with four common antipsychotics: aripiprazole, risperidone, quetiapine, and olanzapine and the group of healthy, unmedicated control subjects.

Table 11.1 Demographic and Clinical Characteristics of the Patients and Healthy Controls Participating in the Antipsychotic Handwriting. Study Shown are Means (with SD).

	n	% Male	Age (yrs)	Dose, mg/day	Dose, mg/day Risp eq[a]	Total PANSS
Aripiprazole	24	68	49.5 (8.1)	19.8 (11.6)	4.9 (2.5)	56.9 (13.7)
Risperidone	40	70	47.4 (9.6)	4.8 (2.8)	4.8 (2.8)	66.7 (17.1)
Quetiapine	14	77	49.6 (6.2)	443.3 (271.7)	4.9 (3.4)	70.1 (17.7)
Olanzapine	13	83	52.4 (6.6)	13.5 (7.3)	4.5 (2.0)	61.0 (21.9)
Controls	57	41	41.9 (9.4)			

[a] Daily dose was scaled in risperidone equivalent dose based on tables published by the expert consensus panel, *Journal of Clinical Psychiatry* 2003; 64 (Suppl. 12).

Table 11.2 Means (and Standard Deviations) for Key Kinematic Parameters Derived from Analysis of All Pen Strokes Recorded during Written Production of the Sentence "Today Is a Nice Day" for Subjects Grouped by Primary Antipsychotic Medication and a Group of Healthy Controls

	Stroke Duration, ms	Stroke Length, cm	Average Stroke Velocity, cm/s	ANJ	No. Acc. Peaks
Aripiprazole	221 (80)	0.58 (0.16)	4.89[b] (2.15)	45.11 (38.09)	1.59 (0.41)
Risperidone	254[a] (91)	0.71 (0.17)	5.42[c] (2.21)	49.42[d] (40.94)	1.85[e] (0.57)
Quetiapine	181 (32)	0.68 (0.18)	6.82 (1.90)	34.34 (20.48)	1.41 (0.22)
Olanzapine	199 (49)	0.70 (0.13)	6.06 (1.46)	41.25 (33.96)	1.51 (0.25)
Controls	172 (47)	0.67 (0.16)	6.80 (1.71)	23.32 (12.95)	1.37 (0.28)

[a] Significantly greater than quetiapine ($p < 0.01$), olanzapine ($p < 0.10$), and healthy controls ($p < 0.0001$).
[b] Significantly lower than healthy controls ($p < 0.01$).
[c] Significantly lower than healthy controls ($p < 0.05$).
[d] Significantly greater than healthy controls.
[e] Significantly greater than quetiapine ($p < 0.01$) and healthy controls ($p < 0.0001$).

Table 11.2 shows the key findings from the analysis of handwriting kinematics for the sentence writing task. Significant group effects were observed for all kinematic variables except stroke length. The finding that our psychosis patients on average did not differ from healthy control subjects on a measure of stroke length suggests that second-generation antipsychotics are not likely to cause parkinsonian micrographia. For the main effects of medication group, the risperidone group exhibited significantly longer stroke durations, lower stroke velocities, and greater stroke dysfluency than healthy controls and, for some variables, than patients treated with olanzapine. The increase in handwriting slowness and dysfluency could not be attributed simply to higher medication dose as patients treated with aripiprazole, quetiapine, or olanzapine received on average the same daily dose (when scaled in risperidone equivalents; see Table 11.1).

This does not suggest that abnormal handwriting kinematics were not dose related, but rather that they may be due to some other property of the antipsychotic. Unlike quetiapine or olanzapine, risperidone has significant dopamine D2 receptor blocking properties. Aripiprazole, on the other hand, is a dual dopamine receptor antagonist (as is risperidone) and agonist (unlike other antipsychotics). This dual mechanism of action appears to protect the patient from some motor effects (dysfluency), but not all (slowness).

Overall, the findings of this study revealed handwriting patterns that seemed to be associated with the dopamine receptor blocking properties of the antipsychotic. Interestingly, while the subtle handwriting motor impairments associated with second-generation antipsychotics appear not to include micrographia, they do include other handwriting disturbances such as increased slowness, increased dysfluency, and reduced stroke duration. These findings suggest that drug-induced parkinsonism may be distinguished from idiopathic Parkinson's disease on the basis of a handwriting kinematic profile.

Performance on several handwriting kinematic variables correlated with the daily equivalent dose and type of antipsychotic medication. A dose of aripiprazole was associated with slowing and more dysfluencies of the movement as expressed by an increase in movement duration ($r = 0.70$; $p < 0.05$) and a decrease in smoothness ($r = 0.86$; $p < 0.001$). However, a dose of risperidone was mainly associated with the dysfluency measures such as decreased smoothness ($r = 0.66$; $p < 0.01$) and increased number of acceleration peaks ($r = 0.55$; $p < 0.01$).

We examined handwriting movements in 22 patients on two occasions, separated by an average of 1 month. Fifteen of the patients remained on stable antipsychotic doses for the two assessments; seven underwent dose increase between the first and second assessment. To compare the mean daily dose across groups of patients better, the dose for any given antipsychotic was adjusted using risperidone equivalents (Expert Consensus Panel 2003). That is, by converting the daily dose of aripiprazole, olanzapine, or quetiapine to a standard risperidone equivalent (see Table 11.1), we could describe the group change in antipsychotic dose using a standard metric. The mean (sd) antipsychotic dose for the 15 stable patients was 3.76 (2.63) mg/day risperidone equivalents for both assessments. The mean antipsychotic dose for the seven dose-switching patients before the dose increase was 2.85 (1.95) mg/day risperidone equivalents, which was increased to a mean of 6.14 (2.34) mg/day risperidone equivalents.

Analyses of their handwriting kinematics for the two assessments revealed that for all handwriting tasks involving the dominant hand combined, patients undergoing antipsychotic dose increase exhibited significantly lower peak vertical velocities compared to stable patients. No other kinematic comparisons were significant. These results are shown in Figure 11.2. The findings support the use of handwriting kinematics as a marker of emergent parkinsonism associated with increasing the dose of dopamine-blocking medications.

In summary, from studies of the effect of antipsychotics on handwriting kinematics, differences can be detected across medications and daily doses. Antipsychotics with greater dopamine receptor blocking properties induce greater slowing and less smoothness in handwriting movements than medications with little or no dopamine antagonism. The longitudinal findings supported the ecological validity of handwriting movement analysis as an objective behavioral biomarker for quantifying the effects of antipsychotic medication and dose on the motor system.

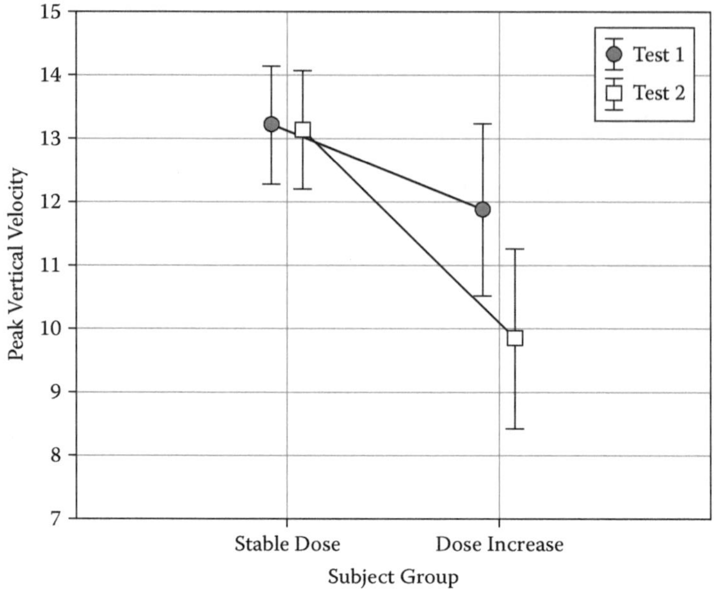

Figure 11.2 Mean peak vertical velocity for medically stable patients and patients undergoing antipsychotic dose increase over two assessments (test 1 and test 2).

Anticholinergics

Anticholinergic medications are prescribed for many medical conditions, including overactive bladder or incontinence, irritable bowel syndrome, pancreatitis, urethral and urinary bladder spasm, respiratory disorders, and parkinsonism. It is not uncommon for patients treated with potent antipsychotics to be prescribed a prophylactic anticholinergic medication (such as benztropine; trade name: Cogentin) to reduce the severity of parkinsonian side effects. Because writer's cramp and dystonia share a common mechanism involving dopamine blockade, benztropine is often prescribed for relief of writer's cramp.

Recall from Chapter 1 that cholinergic interneurons in the striatum synapse on striatopallidal GABAergic neurons and modulate pallidal inhibition. Loss of the effectiveness of the striatal cholinergic interneurons leads to a subsequent reduction in striatopallidal inhibition. It is through this mechanism of reducing striatopallidal inhibition that anticholinergics are effective in counteracting the parkinsonian effect of a dopamine antagonist drug. However, the problem is that anticholinergics can also lead to excessive dyskinetic movements in some patients.

As a proof of concept, we examined data from the larger antipsychotic study (described earlier) to understand the effects of anticholinergic

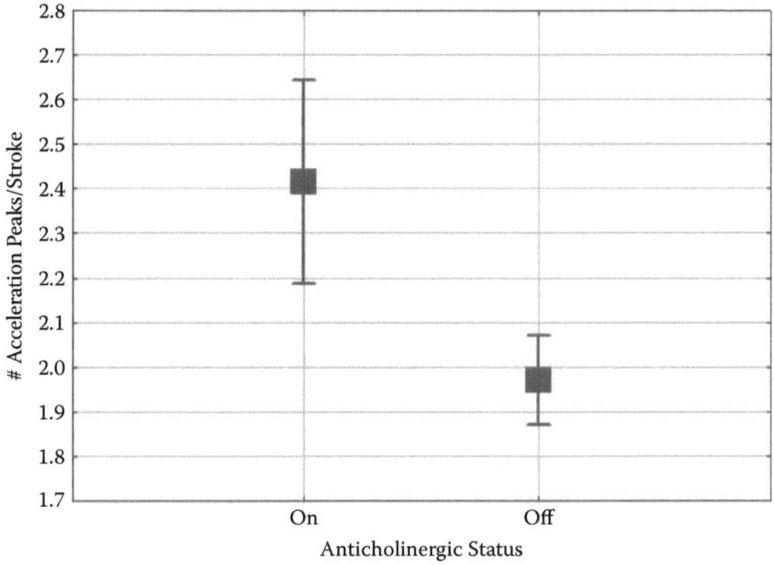

Figure 11.3 Effects of prophylactic anticholinergic medication on smoothness of handwriting movements. Shown are the means (with 95% confidence bars) for the number of acceleration peaks (inversions) during handwriting for patients treated with (n = 21) and without (n =108) anticholinergic medications.

medication on handwriting. Based on the role of acetylcholine as a modulator of nigro–striato–pallidal neurotransmission, we reasoned that blocking acetylcholine with an anticholinergic medication would increase motor activity through a disinhibition mechanism and manifest as a decrease in smoothness of handwriting movements. To test this idea, we examined the number of acceleration peaks per stroke as an index of excessive movement. Of the 129 patients with complete medication information, 21 were treated with an anticholinergic agent. Kinematic results were combined across all strokes for all tasks involving the dominant hand only.

Results revealed a significant main effect of anticholinergic status ($F_{6,874} = 5.12$; $p < 0.001$). This effect is shown in Figure 11.3. With the exception of the sentence task, the number of acceleration peaks per stroke was greater for patients on anticholinergic than off anticholinergic medication. These findings underscore the sensitivity of handwriting kinematics to effects of psychotropic medications that have very specific mechanisms of action.

Antidepressants

Psychomotor disturbances are ubiquitous in depressive disorders. The majority of patients suffering from clinical depression exhibit forms of psychomotor retardation, although psychomotor agitation is not uncommon (Sobin and Sackheim 1997; Schrijvers, Hulstijn, and Sabbe 2008). Research has demonstrated that some aspects of handwriting, particularly handwriting speed and variability for drawing complex figures, are impaired in patients with depression (van Hoof et al. 1993; Sabbe et al. 1999, 2006; Mergl et al. 2004). For example, Mergl et al. (2004) examined handwriting kinematics in patients with clinical depression and healthy control subjects of comparable age. Subjects were asked to draw concentric circles and to write a standard sentence using their natural writing style. All patients were treated with antidepressants at the time of the handwriting assessment.

Several group differences were reported for circle drawing, including variability in peak velocity (depressed > controls) and variation in stroke duration (depressed > controls); however, no differences were observed for mean stroke duration, stroke length, or stroke velocity. Analysis of sentence writing revealed longer stroke durations for depressed than control subjects. Mergl et al. (2004) speculated that as handwriting is a highly automatic motor skill, it is unlikely that depressed patients would exhibit impaired handwriting to a significant degree. On the basis of their work, it is plausible that, unlike figural drawing or handwriting tasks that involve complex psychomotor processes, automatic forms of handwriting such as sentences or signatures may not be noticeably impaired in depressed patients.

Research of the antidepressant effect on measures of fine motor control such as handwriting generally serves two purposes: to identify predictors of response or to better understand pharmacological mechanisms of psychomotor impairment. Earlier research on handwriting changes in patients treated with antidepressants employed quantitative measures of figure drawing rather than handwriting per se (Sabbe et al. 1996, 1997; Mergl et al. 2004; Schrijvers et al. 2009). The rationale was that psychomotor retardation was predominantly a cognitive-motor disturbance and that figure drawing allowed the separation of the cognitive from motor processes underlying psychomotor depression. Sabbe et al. (1996) examined reaction (cognitive) and movement (motor) times associated with figure drawing in patients before and following treatment with fluoxetine (Prozac). They reported that reaction time but not movement time improved following treatment.

In a follow-up study to examine the motor component of figure drawing more closely, Sabbe et al. (1997) tracked changes in hand movement time and velocity following 6 weeks of fluoxetine therapy. While both movement time and velocity improved with therapy, they did not return to normal levels despite the marked clinical improvement of the patients. These findings suggest that slow handwriting movements associated with clinical depression may be resistant to antidepressant therapy.

Schrijvers et al. (2009) examined graphic motor ability in 19 patients prior to and following 6 weeks of therapy with sertraline (an SSRI also considered to be a dopamine uptake inhibitor). Writing tasks included copying lines, simple figures, and complex figures. Pen movements were digitized and their temporal features analyzed. Results revealed significant decrease in initiation and movement times for simple figures and lines, but not for complex stimuli. These findings differ from previous research by Sabbe et al. (1996, 1997) showing improvement in cognitive but not neuromotor processes following fluoxetine therapy. One explanation for the discrepancy is that, unlike fluoxetine, sertraline has both serotonergic and dopaminergic properties. Pharmacological enhancement of dopamine availability within the basal ganglia could be responsible for the reduction in movement time not observed with fluoxetine.

Two interesting studies from Germany compared handwriting movements of patients treated with different classes of antidepressants (Tucha et al. 2002; Hegerl et al. 2005). Tucha et al. (2002) assessed pen movement time, velocity, and acceleration during handwriting of simple sentences for patients treated with a tricyclic antidepressant (TCA) compared to patients treated with an SSRI antidepressant. Handwriting samples were digitized and subjected to computerized analyses of pen stroke characteristics. Results indicated that patients treated with TCAs displayed increased movement times, reduced peak velocity, and reduced acceleration of descending strokes during sentence writing. No kinematic deficiencies were observed for the SSRI-treated patients. Unfortunately, the Tucha et al. study utilized a cross-sectional design comparing two patient groups rather than a longitudinal design to evaluate change due to the antidepressant. Thus, it was not possible to determine whether the TCAs induced the handwriting impairment or whether the TCAs were ineffective in treating psychomotor retardation. Two alternative explanations are possible: Either SSRI treatment did not induce handwriting impairment or SSRIs were more effective than TCAs in treating psychomotor retardation.

In a similarly designed study, Hegerl et al. (2005) examined handwriting kinematics from 16 patients treated with SSRI (citalopram) and compared them with 12 patients treated with a noradrenalin reuptake inhibitor (NARI; reboxetone). Patients were examined prior to treatment and then 4 weeks following treatment. The researchers found that patients treated with SSRIs had significantly reduced stroke movement frequencies during sentence writing and reduced tangential velocities during rapid drawing of circles compared to those treated with reboxetine. Unlike the Tucha et al. study, the Hegerl et al. study did employ a longitudinal study design, so it was possible to draw conclusions about causality. Despite the relatively small sample of patients, the findings demonstrate that SSRIs are more likely to induce subtle handwriting impairment (in the form of slower movements) than NARIs.

The Tucha et al. (2002) and Hegerl et al. (2005) findings have direct relevance to forensic applications. Specifically, handwriting samples of individuals treated with tricyclic antidepressants may show signs of slowness or other motor impairment not likely found in samples from individuals treated with SSRI or NARI antidepressants. Thus, it would be important to document whether the writer was or was not treated with antidepressants as well as the class of antidepressant with which the writer was treated.

Summary

Huber and Headrick (1999) emphasized the importance of accurately discriminating between disguise and the effects of medication on handwriting. Empirical research demonstrates that psychotropic medications alter handwriting in ways that can easily be misinterpreted as disguised. At least two challenges face document examiners when they attempt to discriminate between disguise and genuine handwriting in samples produced by a writer known to have been treated with psychotropic medications. The first is that the illness for which the medication was initially prescribed often presents with a movement disorder affecting fine motor control of the hand. Spontaneous hand dyskinesia (Caligiuri and Lohr 1994) and parkinsonism (Caligiuri and Lohr 1993) are not uncommon in untreated patients with psychosis. The second challenge pertains to the variable effects of the medications on handwriting over time. The time required for patients to develop tolerance to the acute side effects of antipsychotics varies across patients. Also, older patients are more vulnerable to drug-induced motor side effects than younger patients (Caligiuri, Jeste, and Lacro 2000). These considerations underscore the importance of careful documentation of medication and symptom histories for individuals presenting questioned documents.

The goal of this chapter was to explore the various effects of psychotropic medications on handwriting. Over 50 years ago, investigators recognized the importance of assessing handwriting to estimate optimal doses of a neuroleptic and to manage medication intolerance. Despite advances in drug development over the past 20 years and greater access to pharmacotherapies with fewer side effects than previously available medications, subtle drug-induced motor side effects remain a problem for many patients. Using sensitive kinematic procedures to obtain and analyze handwriting samples from hundreds of psychiatric patients, we were able to demonstrate that these newer second-generation antipsychotics can produce subtle forms of handwriting impairment.

While there is an emerging literature on effects of antidepressants on handwriting, similar research for anxiolytics or mood stabilizers used to treat patients with anxiety disorders or bipolar disorder, respectively, is sorely lacking. This is problematic because a significant proportion of patients diagnosed with a psychiatric disorder are treated using combinations of antipsychotics, antidepressants, and anxiolytics. Their synergistic effects on handwriting are presently unknown.

Substance Abuse and Handwriting 12

Introduction

As handwriting may be considered a highly complex motor behavior, it is reasonable to expect that abuse of recreational drugs that alter neuromotor system functions would also impact handwriting. In this chapter, we will explore the effects of commonly abused drugs such as methamphetamine, cannabis, and alcohol on handwriting kinematics. The National Institute on Drug Abuse (NIDA 2009) has been compiling statistics for many years on the prevalence of substance abuse worldwide. Estimates of the economic, societal, and legal costs of substance abuse in the United States exceed $500 billion annually (Nicosia 2009). *Cannabis sativa* has been a part of the human medicinal and cultural experience for over four millennia. Today cannabis is used mainly for recreational purposes because of its euphoric properties. While the epidemiology of cannabis use remains uncertain, it has been estimated that over 160 million adults have used cannabis worldwide (United Nations Office on Drugs and Crime; UNODC 2008), with the highest consumption reported in the Unites States, Australia, and New Zealand (Hall and Degenhardt 2009).

After marijuana, amphetamines are the most widely used illicit drug worldwide (UNODC 2008). There are about 25 million amphetamine users, which exceeds the numbers of cocaine and heroin users combined. In 2005, 39% of state and local law-enforcement agencies cited methamphetamine as their greatest drug threat. The number of individuals aged 12 or older reporting past-year methamphetamine use was approximately 1.3 million in 2007 (National Survey on Drug Abuse and Health [NSDUH], annual survey by the Substance Abuse and Mental Health Services Administration, www.samhsa.gov). It is estimated that 0.2% of the US population currently (in the past month) use methamphetamine. Of the estimated 150,000 people who used methamphetamine for the very first time in 2007, the mean age was 19 compared to 22 in 2006. In 2006, 18- to 25-year-olds were the most likely users of methamphetamine (13%). Growth in amphetamine-related hospital admissions (primarily methamphetamine) increased in each region of the United States between 1992 and 2005.

In this chapter, we provide a general background of the neurobiology of substance abuse with specific reference to methamphetamine, cannabis, and alcohol followed by a summary of the literature on movement disorders associated with these substances of abuse. We then present findings from recent research from our laboratory on handwriting among individuals exposed to methamphetamine or cannabis. Finally, we discuss the implications of this research on forensic applications.

Methamphetamine

Neurobiological Mechanisms Underlying Methamphetamine-Induced Movement Disorders

As noted throughout this book, dopamine is an important neuromodulator active in regions of the brain that control movement, emotion, motivation, and feelings of pleasure (collectively known as reward circuits). Methamphetamine (as well as other recreational drugs such as cocaine and ecstasy) induces a surge in dopamine throughout these reward circuits. As a person continues to abuse these drugs, the brain adapts to the overwhelming surges in dopamine by producing less dopamine or by reducing the number of dopamine receptors in the reward circuit. As a result, dopamine's impact on the reward circuit gradually diminishes, reducing the expected effect of the drug. This decrease compels those addicted to drugs to keep abusing drugs in order to increase dopamine to normal levels. They require increasingly larger dosages to achieve the dopamine high—an effect known as increased tolerance. Drugs affect the dopamine level in two or more ways: (1) by imitating the brain's natural chemical messengers, and/or (2) by overstimulating the "reward circuit" of the brain. Methamphetamine, for example, like cocaine, increases the release and blocks the reuptake of dopamine, leading to high levels of the chemical in the brain.

Chronic exposure to recreational stimulants such as methamphetamine can have neurotoxic effects in brain regions mediating motor control (Ricaurte et al. 2002; Parrott et al. 2002). The primary pharmacological effect of methamphetamine on dopamine is to facilitate presynaptic release of dopamine with secondary effects of inhibiting dopamine reuptake and metabolism (Stahl 1996). According to models of basal ganglia function, increasing nigrostriatal dopamine increases striatopallidal GABAergic inhibition within the indirect pathway. Loss of GABAergic output throughout the basal ganglia leads to an increase in glutamatergic excitation within the thalamocortical pathway and produces excessive movement or dyskinesia. With chronic administration of methamphetamine, neurotoxic effects begin to take place. Ricaurte et al. (2002) and Chapman et al. (2001) have shown

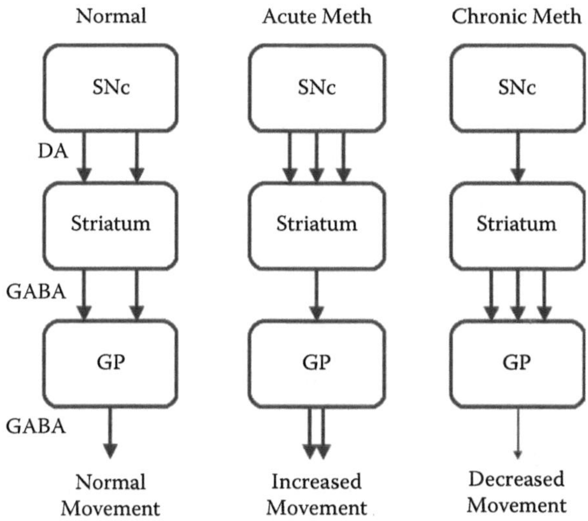

Figure 12.1 Diagram depicting effects of acute and chronic methamphetamine exposure on dopamine (DA) and GABA neurotransmission within the nigrostriatal and striatopallidal pathways, respectively, and their putative behavioral consequences. In acute methamphetamine exposure, excessive nigrostriatal (SNc–striatum) DA reduces striatopallidal (striatum–GP) inhibition and leads to excessive movement. In chronic methamphetamine exposure, loss of nigral cells (SNc) leads to nigrostriatal inhibition and causes excessive striatopallidal inhibition and a reduction in movement. For simplification, important projections from the striatum to the globus pallidus (GP) via the subthalamic nucleus are not shown.

that after the initial increase in dopamine, there follows marked depletion, especially within the striatum.

In general, the loss of striatal dopamine is associated with a reduction in neuropeptide function that normally acts to inhibit GABA. Loss of inhibitory regulation of striatopallidal GABA causes GABA levels to increase, which leads to a reduction in glutamatergic excitation within the thalamocortical pathway and produces parkinsonian-like motor slowing. A diagram depicting a simplified model of the neurotransmitter changes that decrease or increase inhibition is portrayed in Figure 12.1. This model offers an explanation of the neurotransmitter mechanisms thought to be responsible for methamphetamine-induced parkinsonism and hyperkinesia. Contemporary models of basal ganglia function show that dysfunction within the nigrostriatal or striatopallidal circuits produce not only a failure to facilitate desired movements (i.e., parkinsonism or hypokinesia), but also a failure to inhibit unwanted movements (i.e., chorea, hyperkinesia, and tics) (Mink 2003).

Methamphetamine exerts powerful influences on brain systems regulating cognitive and sensorimotor functions. The literature includes reports on three forms of disordered movement associated with methamphetamine abuse: persistent parkinsonism, acute hyperkinesia, and psychomotor disturbances. (For a review of this literature, see Caligiuri and Buitenhuys 2005.) Collectively, these studies suggest that movement disorders stemming from changes to dopaminergic neurotransmission in the basal ganglia likely originate from terminal degeneration at the neuronal level and/or a compensatory homeostatic response to the neurotoxic effects of methamphetamine (Guilarte 2001). Based on the available preclinical and human literature, Caligiuri and Buitenhuys (2005) hypothesized that persistent irreversible movement disorders appear to implicate a primary degenerative process and may take the form of parkinsonism, whereas acute methamphetamine-induced movement disorders implicate secondary adaptive processes and take the form of hyperkinesia.

While the literature on extrapyramidal motor signs associated with methamphetamine abuse is sparse, studies of psychomotor changes are more abundant. For example, investigators report disturbances on several measures of psychomotor function, such as the grooved pegboard test (Volkow et al. 2001), trail-making tests (Kalechstein, Newton, and Green 2003; Simon et al. 2000), finger tapping (Toomey et al. 2003), and reaction time (Richards et al. 1993; Chang et al. 2002). It is not clear, however, if disturbances on specific psychomotor measures stem from extrapyramidal motor disturbances (e.g., parkinsonism) or reflect nonmotor (e.g., cognitive) disturbances in planning, attention, or executive function.

Using positron emission tomography (PET), Volkow and colleagues (2001) observed a relationship between performance on the grooved pegboard and timed gait tasks and loss of dopamine transporter (DAT) in the striatum in abstinent (12 months) methamphetamine abusers. Slower performance times were associated with lower levels of striatal DAT availability. Kalechstein et al. (2003) reported that abstinent methamphetamine users had significantly poorer performance on measures of psychomotor speed (symbol digit modalities test) compared with controls. Subjects in this study were assessed at least 5 days following a positive urine test for methamphetamine. It is difficult to generalize the results from these two studies with regard to psychomotor impairment because of differences in the tasks used to assess psychomotor speed, lack of adequate control groups, and difference in duration of abstinence. Nonetheless, these findings indicate that psychomotor changes can persist beyond the "crash" phase following methamphetamine use.

Human studies suggest that the psychomotor disturbances associated with methamphetamine may not be due to basic motor processes, but rather involve higher level motor processes such as set shifting, planning, and manipulation of information (Simon et al. 2000). This conclusion is consistent with the findings by Moszczynska et al. (2004) that the neurotoxic effects of prolonged exposure to methamphetamine are more pronounced in the cognitive areas of the striatum (i.e., caudate) than the pure motor areas (i.e., putamen).

While research on the motor effects due to (1) methamphetamine exposure, (2) its crash phase, and (3) beyond the crash phase is still ongoing, systematic study of handwriting movements may reveal differences across these three stages. In the following section, we present evidence that kinematic measures of handwriting may discriminate between individuals who recently were exposed to methamphetamine from individuals who never used methamphetamine.

Effects of Methamphetamine Handwriting Kinematics

It is well known that illicit drugs can also affect handwriting (Gesell 1961; Purtell 1965). Procedures for diagnosing illness or exogenous intoxication based on handwriting samples were suggested by Buquet and Rudler (1987). In a recently completed pilot study,[1] seven individuals (six males and one female) with recent exposure to methamphetamine participated in a handwriting kinematics task. For comparative purposes, previously published normative data from healthy control subjects with self-reported negative histories for substance abuse who performed the same handwriting tasks using the same instrumentation (Caligiuri et al. 2009, 2010) were included in the statistical analyses. Data from 57 control subjects (20 males and 37 females with a mean age of 42.5 ± 9.4 years) were available for this purpose. Table 12.1 shows the exposure characteristics of the seven methamphetamine subjects. While the average length of time since last use was just over 1 month, two subjects tested positive (based on urine toxicology) for methamphetamine on the day of the handwriting assessment.

Handwriting movements were recorded using a commercial digitizing tablet and a noninking pen with a Wacom Intuos4 digitizing tablet (30 × 22.5 cm, RMS accuracy 0.01 cm; sampling rate 200 samples per second). The tablet is attached to a Microsoft Windows laptop computer running MovAlyzeR software. The handwriting battery consisted of four tasks:

Table 12.1 Characteristics of the Seven Methamphetamine Users Enrolled in the Handwriting Study

	Mean (sd)
Age (years)	46.7 (4.5)
Age at first exposure (years)	23.1 (7.4)
Days since last exposure	31.7 (43.8)
Total number of days of use	4,659 (2,209)
Total amount of use (grams)	4,628 (2,319)

- cursive loops written from left to right within a 2 cm vertical boundary
- cursive loops written from left to right within a 4 cm vertical boundary
- a complex cursive loop sequence consisting of alternating *lleellee* written within a 2 cm vertical boundary
- the sentence "Today is a nice day" written within a 2 cm vertical boundary

Each sample was repeated five times. The resultant handwriting traces were visible in real time only to the examiner. Subjects were prevented from viewing the recorded trace to remove deleterious effects of visual feedback on movement speed and smoothness.

Data analysis involved the following procedures. The X and Y coordinates were low-pass filtered at 8 Hz using a sinusoidal transition band of from 3.5 to 12.5 Hz (Teulings et al. 1984). Movements were then segmented into successive up and down strokes using interpolated vertical-velocity zero crossings. For each segmented stroke vertical length, duration, peak vertical velocity, and number of vertical acceleration peaks were calculated. In addition, handwriting smoothness was quantified by calculating the normalized jerk averaged (ANJ) per stroke (Teulings et al. 1997). Normalized jerk is unitless as it is normalized for stroke duration and length. ANJ is calculated using the following formula: $\sqrt{(0.5 \times \Sigma(\text{jerk}(t)^2) \times \text{duration}^5/\text{length}^2)}$. Longer segment durations and lower peak velocities are reflective of slow movements, or bradykinesia, whereas higher ANJ scores and increased number of acceleration peaks per segment are indicative of dysfluent writing movements, or dyskinesia. Handwriting kinematic variables were extracted automatically for each pen stroke.

Two-way analyses of variance (ANOVA) were performed with subject group as one factor (with two levels) and handwriting task condition as the second factor (with four levels). Results indicated that for all kinematic variables, differences across handwriting tasks were statistically significant. We found significant group differences for vertical stroke size ($F = 44.2$; $df = 1,3$; $p < 0.0001$), average normalized jerk ($F = 14.13$; $df = 1,3$; $p < 0.0001$), and number of vertical acceleration peaks per stroke ($F = 62.36$; $df = 1,3$; $p < 0.0001$). There were no significant group differences for stroke duration or for average stroke velocity. Results are depicted in Figures 12.2 through 12.4 for vertical stroke size, average normalized jerk, and number of acceleration peaks per stroke, respectively. Group differences for pen pressure could not be tested because of the difference in the sensitivity of digitizing tablets used by the methamphetamine and comparison subjects and the lack of calibration data for pen pressure.

Interestingly, individuals with recent history of methamphetamine use showed no impairment on temporal measures of handwriting movement such as stroke duration or speed. We observed the greatest impairment on measures of handwriting smoothness—that is, ANJ and the number of acceleration peaks (or acceleration inversions). Average normalized jerk was significantly

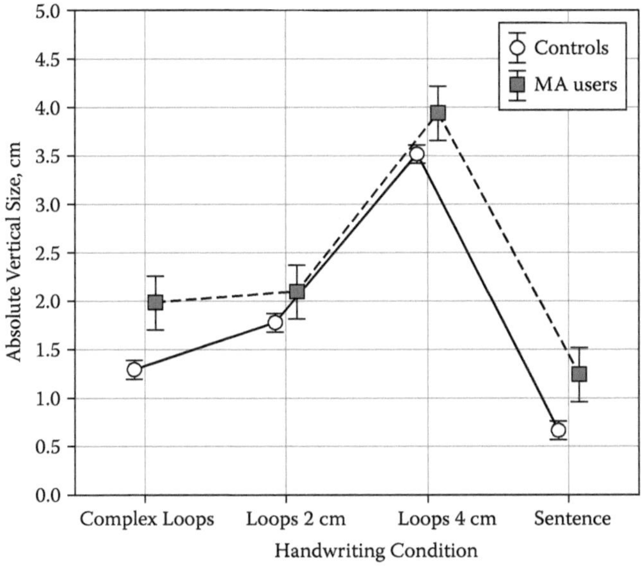

Figure 12.2 Mean (with 95% confidence intervals) vertical stroke length for subjects with histories of methamphetamine abuse (MA users) and nonabuse control subjects for four handwriting tasks. Main effect for the group across all handwriting tasks was statistically significant ($p < 0.001$).

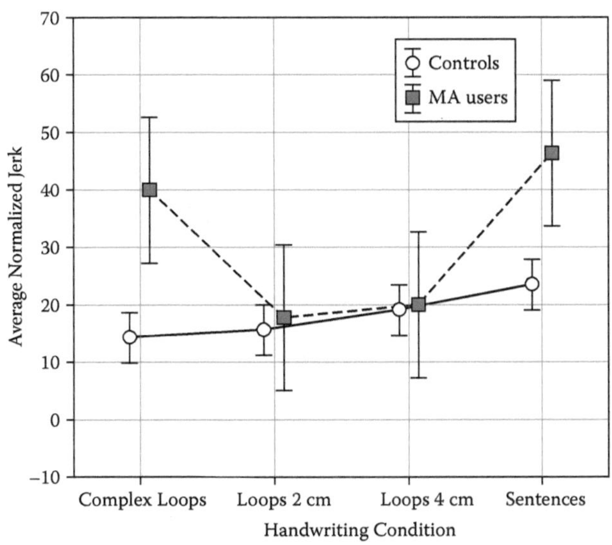

Figure 12.3 Mean (with 95% confidence intervals) average normalized jerk score for subjects with histories of methamphetamine abuse (MA users) and nonabuse control subjects for four handwriting tasks. Main effect for the group across all handwriting tasks was statistically significant ($p < 0.001$).

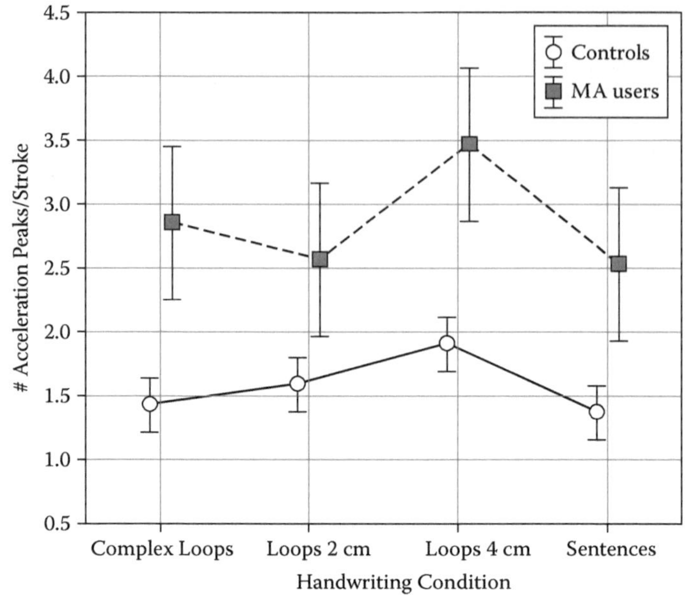

Figure 12.4 Mean (with 95% confidence intervals) number of acceleration peaks per stroke for subjects with histories of methamphetamine abuse (MA users) and nonabuse control subjects for four handwriting tasks. Main effect for the group across all handwriting tasks was statistically significant ($p < 0.00001$).

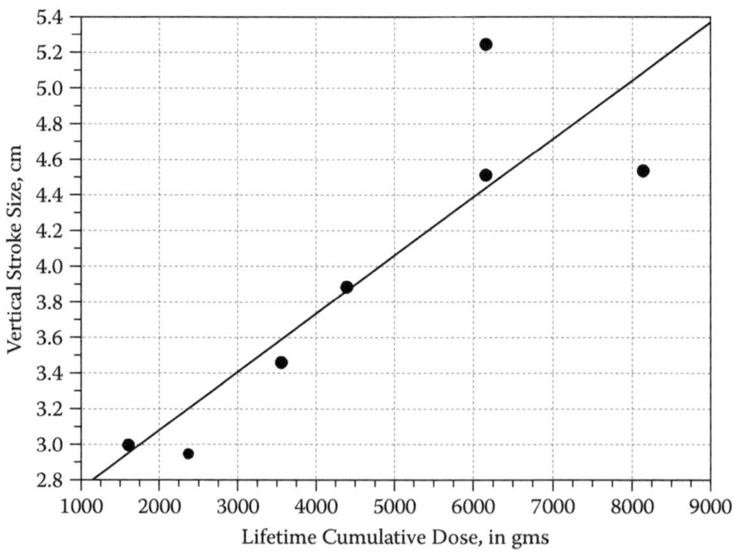

Figure 12.5 Relationship between vertical stroke length (for writing 4 cm cursive loops) and lifetime cumulative dose of methamphetamine.

greater for methamphetamine users than healthy comparison subjects on handwriting tasks that required attention and complex movements such as writing sequences of *lleellee* or sentences. Of equal importance was the finding that stroke size was higher in methamphetamine users than healthy comparison subjects—again, for handwriting tasks that required attention and complex movements such as writing sequences of *lleellee* or sentences.

Further analyses were performed to examine whether there were any associations between handwriting kinematic scores and the individual characteristics of methamphetamine use. Four such associations were observed. Using Pearson correlation coefficients, we observed a significant negative correlation between time since last exposure (in days) and pen pressure for complex loops ($r = -0.79$; $p < 0.05$) and 4 cm loops ($r = -0.79$; $p < 0.05$). A significant positive correlation was observed between total lifetime length of exposure to methamphetamine and the number of acceleration peaks per stroke for 2 cm loops ($r = 0.76$; $p < 0.05$). Lastly, we found a highly significant positive correlation between total lifetime cumulative dose of methamphetamine and vertical stroke size for 4 cm loops ($r = 0.88$; $p < 0.01$), as shown in Figure 12.5.

In summary, the handwriting kinematics of individuals exposed to methamphetamine for an average of 12.4 years differed in two general respects from the general healthy population. First, methamphetamine users wrote with larger vertical amplitudes than comparison subjects. This was particularly evident for complex or alternating letter sequences. Second, the handwriting movements of methamphetamine users were less smooth and more dysfluent than those of healthy comparison subjects.

Overall, alterations in handwriting associated with recent methamphetamine use (within 1 month) did not resemble parkinsonism as predicted by the chronic dopamine toxicity model (Figure 12.1). Rather, the findings of hyperkinetic handwriting patterns are consistent with the acute exposure-dopamine release mechanism. The observation that individuals with histories of methamphetamine abuse exhibit hyperkinetic handwriting is consistent with current models of methamphetamine-induced increased nigrostriatal dopamine. Our handwriting findings are consistent with animal models of methamphetamine-induced nigrostriatal dopamine release and subsequent hyperdopaminergia.

Cannabis

Neurobiological Mechanisms Underlying Cannabis-Induced Movement Disorders

The psychomotor effects of Δ9 tetrahydrocannabinol (Δ9 THC) are mediated by its antagonistic effects on type 1 cannabinoid receptors (CB1) located throughout the motor regions of the brain (Matsuda et al. 1990; Sanudo-Pena,

Tsou, and Walker 1999). Cannabinoid CB1 receptors are involved in cognitive, memory, reward, pain modulation, and motor functions and are found in relatively high density throughout the cerebral cortex, basal ganglia, and cerebellum (Manzanares et al. 2004). Delta-9 THC-mediated receptor antagonism on glutamatergic, GABA-ergic, and dopaminergic neurons in the basal ganglia could lead to motor disturbances ranging from hypokinesia or slowness via reduction in thalamocortical glutamate to hyperkinesia or excessive movements via reduction in striatopallidal GABAergic inhibition.

There exists a modest literature on the effects of cannabis on psychomotor behavior in humans. Messinis et al. (2006) examined psychomotor slowing using the trail-making test in 40 subjects. Subjects were grouped according to recent and chronic users of cannabis. The researchers reported that on this simple measure of motor speed, recent users, but not chronic users demonstrated impairment. However, for the complex task requiring set switching, both groups were equally impaired. This finding suggests that chronic cannabis users may adapt to the deleterious effects of cannabis on movement speed unless the motor task requires cognitive attention.

The impact of cannabis on cognitively demanding motor speed and coordination tasks was replicated by D'Souza et al. (2008) and Fitzgerald, Williams, and Daskalakis (2009). Two other studies shed light on the relationship between psychomotor performance and cannabis dose. Hunault et al. (2009) reported that while some of their cannabis subjects showed no impairment on a reaction time task, there was a linear relationship between increased reaction time and cannabis dose. Roser et al. (2009) reported that cannabis-induced reduction in finger tapping speed correlated with plasma concentrations of Δ9 THC. The findings from published literature indicate a linear dose–response relationship for simple and complex motor behavior. While the majority of cannabis users enrolled in these studies exhibited deleterious psychomotor effects acutely, some showed little or no effect on motor behavior.

To our knowledge, there have been very few studies describing the effects of cannabis on handwriting kinematics. Zaki and Ibraheim (1983) reported findings from an open label study of a small group of cannabis smokers. They found that handwriting in cannabis users was characterized by a decrease in smoothness as evidenced by insertions of excessive acceleration changes, suggesting that cannabis many have a greater effect on the involuntary component of motor control than the more purposive components (such as speed).

As noted before, the controlled studies of psychomotor effects of cannabis reported changes in the speed or latency of motor response (simple reaction time or movement time). However, none of these studies that we are aware of employed handwriting measures and none investigated whether ingestion of cannabis led to hyperkinetic movements or motor disinhibition, as would be predicted by Δ9 THC-mediated GABAergic receptor antagonism

(Fitzgerald et al. 2009). With respect to handwriting movements, evidence of motor slowing or other temporal features associated with cannabis cannot be reliably evaluated using static samples, whereas, signs of hyperkinesia such as the introduction of unwanted movements can be observed and quantified from static handwriting samples. However, given that research on the effects of cannabis on handwriting is sparse, it is important first to evaluate whether cannabis imparts any change to handwriting movements and, if so, to explore the kinematic nature of these changes.

Our laboratory has been fortunate to participate in a recent study of medicinal cannabis for pain management associated with diabetic neuropathy.[2] Five subjects completed a substudy of handwriting kinematics as part of a larger battery of assessments to evaluate therapeutic effects of cannabis on pain associated with diabetic neuropathy. Individuals completed the same procedures as for the methamphetamine study described previously. The study was designed as a within-subject randomized, placebo-controlled trial. Subjects were randomized to four doses: 2.5, 5.0, 7.5 mg Δ9 THC, or placebo. Assessments were conducted 15 minutes prior to dosing (baseline) and 45 minutes following dosing (postdose). Cannabis was inhaled using a vaporizer and subjects were instructed to make 10 inhalations over 5 minutes.

For the purpose of this chapter, we report findings from five subjects who met inclusion criteria and were able to complete pre- and post-dose assessments. Of the five, three received 7.5 mg and two received 2.5 mg Δ9 THC. Single sample t-tests were used to test differences between the baseline and postdose difference score versus zero for each handwriting kinematic variable across the four handwriting tasks.

Figure 12.6 shows examples from one subject for the continuous production of rapid circles. Shown are single trials from the baseline (top) and postdose (bottom) assessment. It should not be surprising that cannabis reduced the frequency of movement from nearly four loops per second to one loop every 2 seconds. However, the introduction of rhythmic micromovements was unexpected and has not been reported in previous cannabis studies. The amplitude of this tremor was too low to resolve as a separate peak in the fast Fourier transform (FFT) plot; however, if one counts the peaks, it is apparent that the frequency of this tremor is consistent with fine high-frequency physiological (postural) rather than coarse pathological tremor, perhaps indicating fatigue. Given that the tremor manifests during handwriting, this would be considered a postural tremor.

Figure 12.7 shows the baseline (predose) and postdose mean scores for stroke duration, stroke length, stroke velocity, and number of acceleration peaks per stroke for sentence writing. Subjects 1, 2, and 5 were administered 7.5 mg Δ9 THC, while subjects 3 and 4 were administered 2.5 mg Δ9 THC.

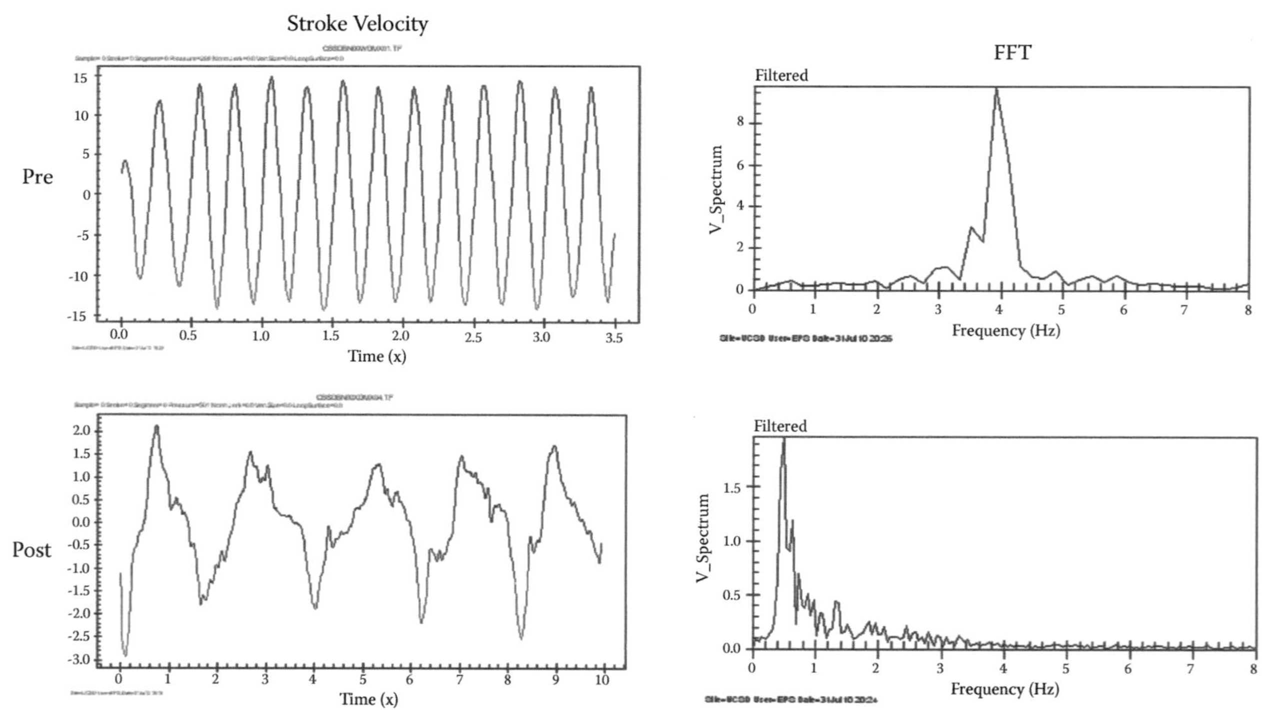

Figure 12.6 Sample velocity (left) and spectral (right) waveforms associated with writing repetitive concentric circles as rapidly as possible for subject CB1 before (top) and after (bottom) exposure to 7.5 mg THC. Of note are the low-amplitude oscillations that appear in the velocity trace, suggesting the presence of postural tremor.

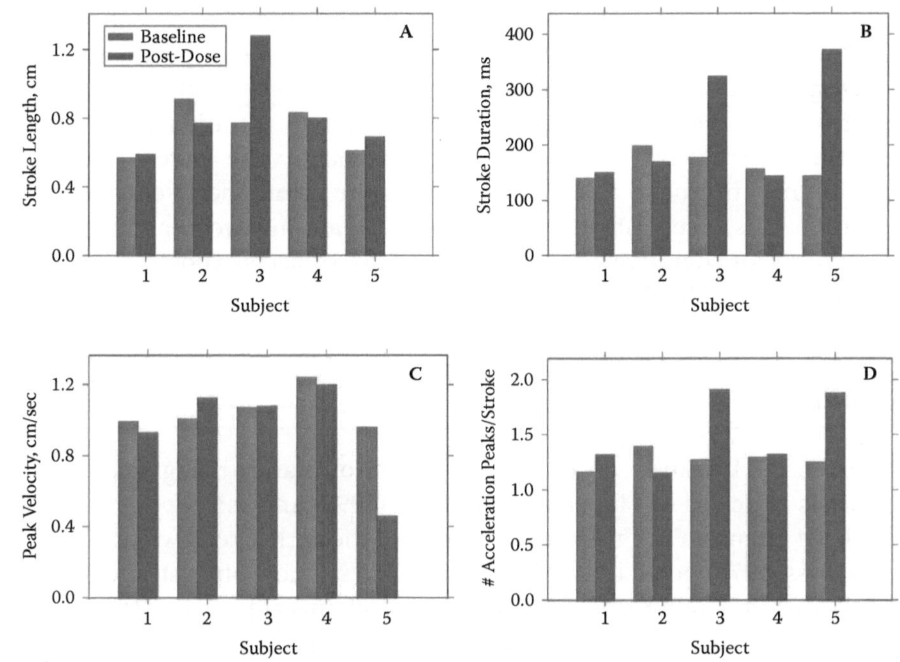

Figure 12.7 Predose baseline (gray bars) and postdose of cannabis (black bars) mean scores for stroke length (A), stroke duration (B), stroke velocity (C), and number of acceleration peaks per stroke (D) for sentence writing.

Results did not reveal a dose effect on handwriting kinematics for these subjects.

Stroke length increased for one subject, while duration increased for two. Stroke velocity was only reduced in subject 5. The number of acceleration peaks increased from baseline to postexposure in subjects 3 and 5, suggesting an increase in handwriting dysfluency. Overall, subject 5 was particularly sensitive to the effects of cannabis. Four of the five subjects exhibited an increase in pen pressure following cannabis exposure (not shown in figure). The mean baseline pen pressure for sentence writing was 635 arbitrary units, which increased to 851 after exposure. Only subject 5 exhibited a decrease (from 789 to 314 units). It is difficult to account for a cannabis-induced mechanism that would cause an increase in pen pressure during writing other than to speculate that perhaps subjects experienced an analgesic effect and, to compensate for weakness in grip strength, exerted greater pressure against the writing tablet.

Interestingly, changes in stroke length were accompanied by proportional changes in stroke velocity for all subjects. This suggests that cannabis appeared not to disrupt the handwriting motor program per se as pen movements adhered to the isochrony principle. When cannabis impacted handwriting kinematics, the disruption appeared limited to execution (general slowness) of movement rather than access to the motor program.

Despite the small sample of subjects, the findings revealed a consistent pattern that inhaled cannabis prolongs movement duration and disrupts smoothness or fluency of handwriting movements. The increase in stroke duration may be due to a delay in executing the command to change stroke direction. This is consistent with previous studies on psychomotor slowing, particularly for cognitively demanding tasks, in subjects exposed to cannabis (Messinis et al. 2006; Roser et al. 2009).

Alcohol

An extensive literature on effects of alcohol on handwriting spanning 50 years was reviewed by Huber and Headrick (1999) and this topic will receive only brief attention in this chapter. In one of the earlier reviews on this subject, Gross (1975) noted that alcohol was the drug most often studied in relation to altered handwriting. Despite multiple claims at that time that alcohol could be responsible for differences in document specimens, very few of the claims were based on empirical research.

One exception was a study by Rabin and Blair (1953), who systematically examined subjective judgments of writing samples and objective measurements. They asked 40 adults to write their signatures and to copy a set of standard words prior to and following the consumption of a "substantially large dose of alcohol." The investigators analyzed several handwriting features including writing speed spatial width, length, size, and accuracy (margin variability) of signatures. Their key finding was that under the influence of alcohol, writers tended to make more errors and require more time and space to complete the writing task than prior to ingesting alcohol. The magnitudes of the temporal and spatial alterations were dose related. Subsequent studies by Tripp, Fluchiger, and Weinberg (1959), Hilton (1969), and Brun and Reisby (1971) confirmed the general finding that, under the influence of alcohol, handwriting movements increase in both size and spatial dimensions (Hilton 1969), become slower, include jerky or broken strokes (Tripp et al. 1959), and exhibit fluctuating pen pressure (Brun and Reisby 1971).

More recent studies involving rigorous scientific methods and experimental controls have been published in the forensic sciences literature (Foley and Miller 1979; Galbraith 1986; Stinson 1997; Asicioglu and Turan 2003). Foley and Miller (1979) compared the effects of cannabis and alcohol on handwriting and found that alcohol was more disruptive to handwriting than cannabis. However, Zaki and Ibraheim (1983) examined the separate effects of cannabis and alcohol on handwriting speed, letter formation, stroke length, and alignment and reported just the opposite. In the absence of sufficient scientific controls, the extent to which cannabis or alcohol exerts similar or different degrees of handwriting impairment remains unknown.

Geller et al. (1991) examined the ability of undergraduate students to judge intoxication accurately on the basis of handwriting samples. Overall these "lay judges" were more accurate in classifying pre- versus postintoxication sentence samples (83.7%) than signatures (67.5%). Judgment accuracy was significantly correlated with blood alcohol concentrations (BAC) of the writers. Judgment accuracy increased to 80% for signatures written by individuals with BAC of 0.15 or higher. These findings suggest that signatures are less susceptible to the effects of alcohol than sentences (possibly due to signatures being overlearned and highly programmed) and underscore an important distinction for the document examiner. Clearly, characteristics of an individual's signature are more likely to be more stable over time than other handwriting samples, particularly when written by an excessive drinker.

Investigators are beginning to apply sensitive quantitative methods to static (Asicioglu and Turan 2003) or dynamic (Phillips, Ogeil, and Muller 2009) handwriting samples to understand further the effects of alcohol on handwriting. Asicioglu and Turan (2003) studied handwriting in 73 individuals before and after the subjects consumed alcohol. Handwriting samples were subjected to analyses with a stereomicroscope, direct and oblique angle lighting, and a video spectral comparator. Direct measurements of stroke length and area were made using a digital caliper. Their findings were consistent with prior research demonstrating that alcohol ingestion induced statistically significant increases in word length, stroke height, spacing between words, and tapered ends, as well as an increase in the angularity and jerkiness of letter formation.

Phillips et al. (2009) evaluated handwriting kinematics in 20 young males. Subjects were administered a dose of alcohol that brought their BAC to 0.048%, a relatively low dose compared to previous studies (e.g., Geller et al. 1991). Subjects were instructed to write a set of four cursive "*l*"s 20 times using a noninking pen. Samples were digitized using a Wacom graphics tablet. Subjects completed the task prior to and 30 minutes after consuming the alcoholic beverage. The investigators subjected the digitized samples to automatic analyses of stroke length, duration, peak velocity, time to peak velocity, the number of zero axis crossings in both velocity and acceleration, and pen pressure. Ballistic pen movements were subjected to fast Fourier analyses to examine peak frequency of pen movement.

The authors reported that writing strokes tended to increase in duration following alcohol ingestion ($p < 0.10$). Moreover, stroke length was positively associated with increased BAC. However, no other kinematic differences before and after alcohol ingestion were observed. One interesting finding to emerge from this study was the observation that following alcohol ingestion, the frequency of ballistic movements tended to concentrate around 4

Hz with a discernable peak in the velocity spectrum. As any peak in the movement spectra is indicative of tremor, the authors attributed their finding to the emergence of an action tremor. Action tremors with a frequency between 3 and 5 Hz are common in cerebellar disorders (see Chapter 4) and other conditions affecting cerebellar function such as alcohol intoxication (Marsden et al. 1977; Volkow et al. 2006).

It is likely that the lack of statistically significant differences on their kinematic measures was due to the low levels of alcohol intoxication (0.048% in their study). Indeed, as noted by Geller et al. (1991), judges could not distinguish sober from intoxicated signatures with sufficient levels of accuracy in cases with BAC below 0.15%.

Several mechanisms have been proposed to account for the handwriting and motor changes associated with alcohol. With regard to the general motor impairment, Frye and Breese (1982) and Hanchar et al. (2005) found that alcohol enhanced tonic GABAergic inhibition, leading to depressed neuronal activity within cerebellar granular neurons. Suppression of cerebellar function would disrupt motor behaviors requiring precise timing and synchronization of multijoint movements in a coordinated manner, such as handwriting.

Tiplady et al. (2005), on the other hand, demonstrated that alcohol-induced increase in handwriting size might be explained by selective impairment of kinesthetic feedback. By isolating the effects of visual and kinesthetic proprioception (i.e., sensation of muscle length), Tiplady and colleagues found that the handwriting alterations were greater following manipulation of kinesthetic than of visual feedback. They noted that alcohol reduced the size of the perceived kinesthetic distance, leading to larger movements. Their explanation is consistent with cerebellar hypotheses of alcohol-induced motor effects insofar as the afferent neurons from the muscle receptors (spindles) project onto cerebellar nuclei (Proske and Gandevia 2009).

Summary

Table 12.2 summarizes the effects of various substances of abuse on handwriting. Data for methamphetamine and cannabis are from work conducted in our laboratory. Findings for alcohol are from the published literature. It is interesting that methamphetamine, cannabis, and alcohol impart similar effects on handwriting kinematics, including increases in stroke length, slowness or increased stroke duration, and decreased smoothness. This observation suggests that methamphetamine, cannabis, and alcohol may share a common mechanism of action likely involving the basal ganglia and contributing to motor disinhibition. This is not unexpected given the disinhibitory effects these three substances have on cognitive, emotional, and psychosocial behavior.

Table 12.2 Handwriting Effects of Various Substances of Abuse

Substance	Effects on Handwriting
Methamphetamine	Increase vertical stroke length
	Decrease stroke smoothness
Cannabis	Increase stroke duration
	Increase stroke length
	Decrease stroke smoothness
Alcohol	Increase stroke length
	Decrease writing speed
	Increase pen pressure variability
	Decrease stroke smoothness

From the perspective of the forensic document examiner, samples written while under the influence of methamphetamine or alcohol should exhibit signs that are readily observed by close examination. Unfortunately, increased stroke amplitude and decreased smoothness are not difficult to simulate should someone be motivated to disguise his or her signature. Based on kinematic analyses of dynamic handwriting, it is possible that information such as stroke length and smoothness obtained from the static signature could inform the examiner of substance intoxication.

Notes

1. Research supported by NIDA P30-MH62512 and P01-DA12065.
2. Funding for this study was provided by the state of California under Proposition 63: Medicinal Marijuana Initiative, awarded to Dr. Mark Wallace, University of California, San Diego.

Aging and Handwriting 13

Introduction

As discussed throughout Chapter 6, the gradual decline in dopamine neurotransmission that accompanies advanced age inevitably leads to declining motor function. Handwriting is not likely to be spared by this process. From the qualitative perspective, deteriorating handwriting takes many forms, including uneven line quality and erratic movements (Hilton 1977; Owens 1990). Document examiners called upon to distinguish a genuine from a forged signature of an elderly person are forced to consider the question of age-related deterioration and whether the available exemplars reliably capture the natural effects of aging of the original writer.

Because many factors can contribute to variability in the quality of a handwritten signature, the document examiner likely approaches this challenge by a process of elimination. Armed with information about how the natural aging process impacts handwriting and signature formation, the examiner can face this challenge with less uncertainty. The goal of this chapter is to provide the document examiner with insight derived from empirical research to enable an informed approach to the problem of aging and signature authentication.

Empirical Research on Effects of Aging on Handwriting

There are surprisingly few published studies on handwriting in aging adults. Research on the effects of advanced age on handwriting consists of basic studies of handwriting speed (Dixon, Kurzman, and Friesen 1993; Rodriguez-Aranda 2003), quantitative analyses of handwriting kinematics (Walton 1997), and the utilization of visual feedback that may account for change in handwriting with age (Slavin, Phillips, and Bradshaw 1996; Contreras-Vidal et al. 2002; Smyth and Silvers 1987; Teulings et al. 2002).

Rodriguez-Aranda studied handwriting speed in 155 subjects ranging in age from 22 to 88 as part of a larger study on psychomotor changes in aging. Her findings on handwriting are consistent with the general expectation that

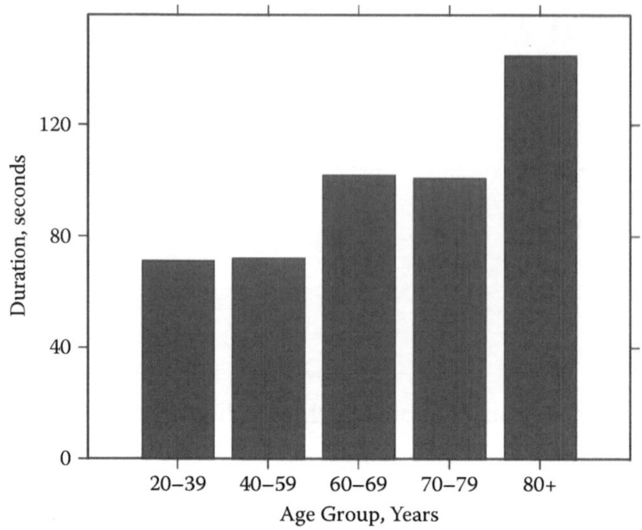

Figure 13.1 Mean duration (in seconds) needed to write 157 characters across five age groups. Main effect of age on handwriting duration was statistically significant ($p < 0.001$). (Adapted from Table 2, page 207, of Rodriguez-Aranda, C. 2003. *Clinical Neuropsychologist* 17:203–215.)

between the ages of 20 and perhaps 70, natural aging effects on handwriting speed are subtle, whereas more noticeable effects are observed after age 70. Figure 13.1 shows the mean durations to complete the handwriting task of 157 characters for subjects within each of five age groups studied. The age-related pattern of motor slowing appears to have a punctuated rather than a gradual pattern; the first increase in writing time occurs after age 60 and then again after age 80.

Walton (1997) evaluated up to 26 features from sentences written by 51 healthy subjects between the ages of 39 and 91, many of whom were reexamined 5 years later. Walton was able to report that among those subjects under the age of 65, handwriting characteristics remained relatively stable over a 5-year period, showing no age-related decline. The most prevalent feature that distinguished middle-aged (39–65) from older subjects was the stroke pattern for pen pressure. Younger subjects produced downstrokes with greater pen pressure (thicker lines) than upstrokes, whereas older subjects showed more uniform pen pressures between upstrokes and downstrokes. There was an association between age and the number of pen lifts such that the youngest writers (mean age of 22) lifted the pen on average 2.7 times within a sentence, whereas the oldest writers (mean age of 75) lifted the pen on average 6.0 times. Lastly, handwriting samples from approximately 20% of the older subjects showed evidence of very mild tremor.

The pen pressure finding is consistent with findings from our laboratory using quantitative methods (see Chapter 9) showing greater pen pressure for downstrokes compared to upstrokes (14% greater) among younger writers. Even younger writers attempting to forge a signature retain this difference, albeit somewhat lower (11%). The Walton finding that pressure differences for upstrokes and downstrokes diminish in the elderly writer has implications for the forensic document examiner. Specifically, handwriting (or signature) samples by younger writers attempting to simulate handwriting of older individuals (e.g., older than age 65) are not likely to show the uniform stroke direction pattern in pen pressure—a clue to the existence of a potential forgery.

Prior studies on handwriting in older adults have examined the role of visual feedback (Slavin et al. 1996; Contreras-Vidal et al. 2002; Smyth and Silvers 1987; Teulings et al. 2002). Use of feedback is an important consideration because, as humans age, they become more reliant upon feedback (especially visual) for accurate motor control while at the same time there is decline in the acuity necessary to process visual information (Bloesch and Abrams 2010; Anderson and Ni 2008; McNay and Willingham 1998). Slavin et al. (1996) reasoned that slowness or hesitancy in handwriting of older adults could reflect greater dependence upon visual feedback to compensate for increased "neural noise." They examined consistency of handwriting under varying conditions of visual feedback (noninking versus inking pen, use of lined versus plain paper, and having participants wear goggles that blocked the lower half of the visual field) presented in counterbalanced order. Stroke consistency, defined as the ratio of the mean divided by the standard deviation, served as the dependent variable.

Slavin and colleagues reported that while stroke duration was significantly longer for older (mean duration of 328 ms) than younger (mean duration of 281 ms) subjects, older subjects' performance under conditions of visual feedback was characterized by increased variability for several kinematic variables, including stroke duration, peak stroke velocity, and time to peak velocity. For example, use of lined paper increased variability for older but not younger writers compared to use of unlined paper. Thus, older subjects' greater reliance on visual feedback led to a decrease in handwriting efficiency. The authors concluded that the slower handwriting movements of older adults may not necessarily stem from their need to reduce error. Instead, they hypothesized that slower handwriting movements may be related to inefficient use of visual information.

While most studies show that older writers do not completely adapt to experimental manipulations of the visual feedback (e.g., gain change or distortion) during handwriting (Ghilardi et al. 2000; Teulings et al. 2002), there is some evidence that adaptation can occur, albeit more slowly, in older writers (Contreras-Vidal et al. 2002). Contreras-Vidal and colleagues reset the gain of the visual display that subjects relied upon to perform name and sentence writing tasks. Vertical gain was decreased to 70% or increased to 140%, thus creating a mismatch between the planned movement and perceived outcome. Subjects were unaware of when the gain manipulation was deployed throughout the series of trials. Vertical and horizontal stroke length, duration, and normalized jerk were computed for each trial.

Results indicated that both younger (mean age of 23) and older (mean age of 70) adults gradually adapted their visuomotor maps to the gain manipulations. These findings suggest that older writers can make effective use of visual feedback to guide handwriting, raising the question of whether slow handwriting movement in the elderly is due to deficits in motor drive, inefficient use of visual and/or proprioceptive feedback, or combinations of the two. Further research in this area is needed to reconcile this question.

A recent study by Woch, Plamondon, and O'Reilly (2011) tested whether older writers adhere to the minimization principle of response optimization despite their age-related neuromuscular slowing. Woch et al. remind us that as people age, movements become slower and less coordinated. However, it is not known whether these decrements are the result of deterioration of the neuromuscular system or failure to utilize compensatory strategies. The investigators utilized predictions from kinematic theory (see Chapter 3) to predict that aging would be associated with an increase in the timing parameters as reflected in a delta-lognormal model. This model allows for separation of two phases of movement execution: 1) the planning phase; and 2) the neuromuscular response phase based on analysis of the velocity profile

Older (ranging in age from 63 to 70) and younger (ranging in age from 26 to 29) subjects were asked to produce bidirectional strokes as rapidly as possible using a stylus. Handwriting strokes were digitized and subjected to a series of complex analyses of the velocity profiles. Results indicated a robust association between age and the time delay of overall response (i.e., motor planning) as well as the timing relations between agonist (go) and antagonist (stop) neuromuscular systems. The timing relations were derived from the velocity profile, which includes two peaks for bidirectional movements. The initial response delay observed for the older subjects suggests impairment of central nervous system properties, whereas the neuromuscular delays reflect impairment in the execution of optimal movement.

Woch et al. concluded that, rather than being additive, the process underlying these time delays was multiplicative in nature. That is, total overall motor slowing of these bidirectional handwriting movements was greater than the sum of the initial response delay (reaction time) and slowing of the neuromuscular response (as measured by the agonist–antagonist synergy). Unlike previous research demonstrating that handwriting movements become slow with age, the Woch et al. study sheds light on the underlying mechanisms responsible for the age-related change in handwriting.

Summary

Perhaps one of the more challenging tasks confronting document examiners is to confirm the authenticity of a set of signatures written by an individual spanning several decades of adult life. The examiner must employ objective criteria to account for age-related deterioration and evaluate whether the available exemplars reliably capture the natural effects of aging of the original writer. The goal of this chapter was to provide the examiner with the necessary background to formulate these criteria. Key to this process is knowledge of how the natural aging process impacts handwriting and signature formation.

The age-related decline in handwriting is not linear. Studies generally show little or no change in the temporal and spatial attributes of handwriting until after age 80. In addition to writing more slowly, elderly writers tend to produce signatures and sentences with greater variability in temporal and spatial stroke parameters.

As noted before, there are surprisingly few published studies on effects of natural aging on handwriting. This research has focused on the effects of age on the utilization of sensory information to guide handwriting movements, recognizing that both sensory and motor processes deteriorate with advanced age. While research on aging and handwriting is limited, this dearth is more than offset by abundant literature on degenerative neurological disease and handwriting conditions (Chapter 10) that typically emerge late in life. Overlapping processes do not necessarily lessen the challenge for the document examiner. Rather, caution should be exercised when drawing conclusions from evidence of deteriorated handwriting that the source of this deterioration is solely age related.

Conclusions

14

The material covered within the pages of this book establishes a foundation for the construction of a scientific framework to support opinion regarding handwriting and signature authentication. Recent judicial challenges to expert testimony now demand that scientific testimony must be based on evidence that is grounded in empirical research. The goal of this book was to integrate the extensive research on neural processes underlying normal and pathological handwriting and how disease, medication, and age alter these processes.

The empirical research and clinical observations summarized in Part 1 of this book inform the understanding of the neurobiology of normal and pathological handwriting. First, with regard to the anatomical bases underlying neural control of handwriting movements, convergent findings from lesion studies, neurosurgical procedures, and functional neuroimaging research support the existence of a complex network of cortical and subcortical regions that govern handwriting movements. This network has at least five cortical zones dominated by the superior parietal lobe (SPL) and the supplementary motor area (SMA). Case reports of patients surviving vascular accidents involving the basal ganglia confirm the importance of the striatum (especially the putamen) in the ongoing monitoring of handwriting movements. Such individuals exhibit impairments in handwriting that resemble Parkinson's disease (PD). However, unlike PD, micrographic handwriting following a basal ganglia stroke is transient, usually disappearing within weeks following the stroke.

There is compelling evidence supporting the existence of multiple parallel cortical-subcortical circuits that function in regulating fine motor control. Important brain areas involved in the handwriting motor circuit include the superior parietal lobe (SPL), the basal ganglia (consisting of the striatum and globus pallidus), thalamus, and SMA. Evidence presented in Chapter 10 on the effects of deep brain stimulation on handwriting supports a role for the subthalamic nucleus in handwriting motor control. The SMA is thought to function as a comparator in this sensory-motor feedback loop. If the SMA is involved in motor tasks requiring internal monitoring as this circuit would suggest, one could hypothesize that activity within the SMA would differ when a writer is producing a forged or simulated signature requiring greater ongoing monitoring than when producing a genuine signature requiring little or no online monitoring.

Computational and cognitive models of motor control are useful for conceptualizing complex systems, such as handwriting. We addressed the long-standing controversy over whether handwriting movements are programmed and, if so, whether the program is hard-wired or flexible. Indeed, the most compelling evidence for the existence of a flexible and adaptive generalized motor program comes from empirical research on handwriting.

Nonetheless, consideration of handwriting as programmed motor behavior can be somewhat problematic. For one, it can be readily observed that handwriting is a serial motor behavior with individual letters making up words and words making up sentences in series. However, the existence of a motor program presumes that the movement parameters for handwriting are not stored as discrete instructions to specific muscles, but rather as a general spatial code representing the final motor output attainable under a variety of physical or environmental constraints.

The ability of a writer to anticipate abrupt changes in the writing surface or writing instrument and evidence of motor equivalence provides strong support for handwriting as a highly flexible motor program. Researchers have demonstrated that handwriting movements subjected to various computational analyses are executed using stoke trajectories that are cost efficient. Efficient movement trajectories are those where jerk is minimized (i.e., reduced number of acceleration changes), movement time is constant despite changes in stroke length (the isochrony principle), and movement velocity is determined by movement curvature. These parametric rules simplify the demands of the motor program and allow greater flexibility and adaptation to environmental constraints.

Based on these three mathematical concepts, one would hypothesize that during natural signature production, the writer exhibits stroke parameters that adhere to a cost minimization principle, whereas in a forgery or disguised signature, the writer is likely to exhibit movement trajectories that are inefficient. We applied the isochrony principle in a kinematic study of forged signatures to test whether a forgery could be distinguished from a genuine signature solely on the basis of the relationship between stroke length and stroke velocity. We reasoned that if handwriting movements adhere to principles of minimization of effort and are programmed to ensure efficiency, nonprogrammed movements such as forgeries would violate these principles. That is, a forged signature is not likely to be learned or produced with kinematic efficiency.

Indeed, this was what we observed. Our findings demonstrated that genuine signatures were produced with a tightly coupled stroke length–velocity relationship, whereas forged signatures exhibited only a weak relationship. To our knowledge, this was the first demonstration that a fundamental principle of motor control could be applied to the study of signature authenticity.

One major source of variation in handwriting over time, particularly among older writers, is the effect of progressive disease. Diseases of the basal ganglia disrupt regulatory control of movement and reduce (as in Parkinson's disease) or exaggerate (as in Huntington's disease) handwriting movements. Research on handwriting movements among individuals with neurological disease can inform underlying pathological mechanisms responsible for the disease and can provide a record of change in disease progress or benefits of treatment. While the time course and clinical management differs for these conditions, there is overlap in their neurochemistry and pathophysiology, particularly with regard to subcortical brain regions that govern motor control. Overlapping pathophysiology suggests that handwriting movements across various disease states would also share common features.

For example, while motor control deficits exhibited by Alzheimer's disease (AD) patients are characterized by higher level psychomotor abnormalities with relatively normal handwriting (adjusted for age), some AD patients exhibit parkinsonian-like movements. These patients are given the provisional diagnosis of dementia with Lewy bodies (DLB) because they share a common neuropathological finding with Parkinson's disease. DLB patients exhibit the same cognitive and behavioral declines as in typical AD with the additional problem of parkinsonism. Using quantitative kinematic analyses, we were able to demonstrate that handwriting movements in DLB differ from those in AD. While AD patients were more variable as a group than healthy writers, they did not differ on measures of handwriting kinematics from healthy writers as a group. However, handwriting movements for the DLB patients resembled those typically observed in PD. Specifically, DLB patients wrote sentences with slower movement velocities, longer stroke durations, decreased stroke length, and an increased number of acceleration inversions.

A significant proportion of the book was devoted to how medications and drugs alter brain systems governing motor control and the consequences of these effects on handwriting. While psychotropic medications offer therapeutic relief for a number of emotional, mood, and behavioral disorders, they are known to produce a wide range of undesirable motor side effects. Given the ubiquitous access of psychotropic medications today, particularly in the aging population, it is important that the forensic document examiner gain an appreciation of the potential influence of these common medications on handwriting. The time course and nature of psychotropic-induced motor side effects are important when evaluating handwriting samples that appear to reflect change in an individual known to have been treated with an antipsychotic agent. Many elderly individuals treated with psychotropic medications for any number of reasons develop parkinsonian side effects. Handwriting for these individuals would be characterized by many of the same features observed in PD, such as micrographia, increased stroke duration, reduced stroke velocity, and possibly tremor.

At least two challenges face document examiners when they attempt to discriminate between disguise and genuine handwriting in samples produced by a writer known to be treated with psychotropic medications. The first is that the illness for which the medication was initially prescribed often presents with a movement disorder affecting fine motor control of the hand, such as dyskinesia or parkinsonism. The second challenge pertains to the variable effects of the medications on handwriting over time. These considerations underscore the importance of careful documentation of medication and symptom histories for individuals presenting questioned documents.

Over 50 years ago, investigators recognized the value of assessing handwriting in managing the therapeutic and countertherapeutic effects of antipsychotics. Despite advances in drug development over the past 20 years and greater access to pharmacotherapies having fewer side effects than previously available medications, subtle drug-induced motor side effects remain a problem for many patients. Using sensitive kinematic procedures to obtain and analyze handwriting samples from hundreds of psychiatric patients, we were able to demonstrate that these newer second-generation antipsychotics can produce subtle forms of handwriting impairment.

While there is an emerging literature on effects of antidepressants on handwriting, similar research for anxiolytics or mood stabilizers used to treat patients with anxiety disorders or bipolar disorder, respectively, is sorely lacking. This is problematic because a significant proportion of patients diagnosed with a psychiatric disorder are treated using combinations of antipsychotics, antidepressants, and anxiolytics. Their synergistic effects on handwriting are presently unknown.

We were fortunate to be able to include new findings from our laboratory on effects of methamphetamine and cannabis on handwriting. It is interesting that methamphetamine, cannabis, and alcohol impart similar effects on handwriting kinematics, including increased stroke length, reduced stoke velocity, increased stroke duration, and decreased smoothness. This observation suggests that methamphetamine, cannabis, and alcohol may share a common mechanism of action likely involving a basal ganglia feedback circuit and contributing to motor disinhibition. This is not unexpected given the disinhibitory effects these three substances have on cognitive, emotional, and psychosocial behavior. Based on kinematic analyses of dynamic handwriting, it is possible that information such as stroke length and smoothness obtained from the static signature could inform the examiner of substance intoxication.

We addressed an important problem facing forensic document examiners: the problem of aging. Perhaps one of the more challenging tasks confronting document examiners is to confirm the authenticity of a set of signatures written by an individual spanning several decades of adult life. The examiner must employ objective criteria to account for age-related deterioration and evaluate whether the exemplars available reliably capture the natural effects of aging of the original writer.

A general finding emerging from the literature on aging is that age-related declines in handwriting are not linear. Studies generally show little or no change in the temporal and spatial attributes of handwriting until after age 80. In addition to writing more slowly, elderly writers tend to produce signatures and sentences with greater variability in temporal and spatial stroke parameters. Advanced age compromises one's ability to organize the inherent kinematic variability optimally and execute a desired movement sequence.

With regard to handwriting motor control, certain age-related impairments will have a more deleterious effect than others. Tremor will clearly impact handwriting movements and reveal stroke dysfluencies and oscillations. A writer's effort to inhibit tremor by increasing muscle stiffness will result in restricted movements and reduced stroke amplitudes. Reduced grip strength will alter both qualitative and quantitative aspects of handwriting and can be readily observed from the pressure traces embedded in paper documents.

The problem of variability is of particular significance to the document examiner. Fluctuations in force steadiness and inconsistent deployment of adaptive strategies can introduce variability in many features of the handwriting movement, including amplitude, slant, smoothness, and pen pressure. More importantly, these fluctuations can occur within a single document and over time between documents.

Conclusions drawn from the empirical research summarized in this book can inform the questions posed by expert document examiners and can guide future research in understanding the source of variability and nature of judgment error in document examination. Our aim was not to present the definitive work on the neurobiology of normal and pathological handwriting, but rather to propose new questions leading to testable hypotheses and to open new doors to the scientific process and understanding of signature and handwriting authentication.

Bibliography

Abbs, J. H., and Winstein, C. J. 1990. Functional contributions of rapid and automatic sensory-based adjustments to motor output. In *Attention and performance XIII*, ed. M. Jeannerod, 627–652. Hillsdale, NJ, Lawrence Erlbaum Associates.

Abend, W., Bizzi, E., and Morasso, P. 1982. Human arm trajectory formation. *Brain* 105:331–348.

Abernethy, B., and Sparrow, W. A. 1992. The rise and fall of dominant paradigms in motor behavior research. In *Approaches to the study of motor control and learning,* ed. J. J. Summers, 3–45. Amsterdam: North–Holland, Elsevier Science Publishers.

Adams, J. A. 1971. A closed-loop theory of motor learning. *Journal of Motor Behavior* 3:111–150.

Ahmed, Z., Josepha, K. A., Gonzalez, J., DelleDonne, A., and Dickson, D. W. 2008. Clinical and neuropathologic features of progressive supranuclear palsy with severe pallido–nigro–luysial degeneration and axonal dystrophy. *Brain* 131:460–472.

Albin, R. L., Young, A. B., and Penney, J. B. 1989. The functional anatomy of basal ganglia disorders. *Trends Neuroscience 12*, 366–375.

Alewijnse, L.C., van den Heuvel, C.E., Stoel, R, and Franke, K. 2009. Analysis of signature complexity. Paper presented to the 14th Conference of the International Graphonomics Society.

Alexander, G. E., Crutcher, M. D., and DeLong, M. R. 1990. Basal ganglia-thalamocortical circuits: Parallel substrates for motor, oculomotor, "prefrontal" and "limbic" functions. *Progressive Brain Research* 85:119–146.

Alexander, G. E., DeLong, M. R., and Strick, P. L. 1986. Parallel organization of functionally segregated circuits linking basal ganglia and cortex. *Annual Review of Neuroscience* 9:357–381.

Alexander, M. P., Fischer, R. S., and Friedman, R. 1992. Lesion, localization in apractic agraphia. *Archives of Neurology* 49:246–251.

Alkahtani, A., and Platt, A. 2009. Relative difficulty of freehand simulation of four proportional elements in Arabic signatures. *Journal of American Society of Questioned Document Examiners* 12:69–75.

Alkhani, A., and Lozano, A. M. 2001. Pallidotomy for Parkinson's disease: A review of contemporary literature. *Journal of Neurosurgery* 94:43–49.

Andersen, K., Launer, L. J., Dewey, M. E., Letenneur, L., Ott, A., Copeland, J. R. M., Dartigues, J.-F. et al. 1999. Gender differences in the incidence of AD and vascular dementia: The EURODEM studies. EURODEM Incidence Research Group. *Neurology* 53:1992–1997.

Anderson, G. J., and Ni, R. 2008. Aging and visual processing: Declines in spatial not temporal integration. *Vision Research* 48:109–118.

Anderson, M. E., and Horak, F. B. 1985. Influence of the globus pallidus on arm movements in monkeys. III. Timing of movement-related information. *Journal of Neurophysiology* 54:433–448.

Apps, R., and Garwicz, M. 2005. Anatomical and physiological foundations of cerebellar information processing. *Nature Reviews Neuroscience* 6:297–311.

Ardila, A., and Rosselli, M. 1993. Spatial agraphia. *Brain and Cognition 22*:137–147.

Arvanitakis, Z., and Wszolek, Z. K. 2001. Recent advances in the understanding of tau protein and movement disorders. *Current Opinion Neurology* 14:491–497.

Asicioglu, F., and Turan. N. 2003. Handwriting changes under the effect of alcohol. *Forensic Sciences International* 132:201–210.

Atkeson, C. G., and Hollerbach, J. M. 1985. Kinematic features of unrestrained arm movements. *Journal of Neuroscience* 5:2318–2330.

Auerbach, S. H., and Alexander, M. P. 1981. Pure agraphia and unilateral optic ataxia associated with a left superior parietal lobule lesion. *Journal of Neurology, Neurosurgery, and Psychiatry* 44:430–432.

Bain, P. G., Findley, L. J., Britton, T. C., Rothwell, J. C., Gresty, M. A., Thompson, P. D., and Marsden, C. D. 1995. Primary writing tremor. *Brain* 118:1461–1472.

Bannon, M. J., Poosch, M. S., Xia, Y., Goebel, D. J., Cassin, B., and Kopatos, G. 1992. Dopamine transporter mRNA content in human substantia nigra decreases precipitously with age. *Proceedings of National Academy of Sciences* 89:7095–7099.

Barbarulo, A. M., Grossi, D., Merola, S., Conson, M., and Trojano, L. 2007. On the genesis of unilateral micrographia of the progressive type. *Neuropsychology* 45:1685–1696.

Barretta, S., Sachs, Z., and Graybiel, A. M. 1999. Cortically driven Fos induction in the striatum is amplified by local dopamine D2-class receptor blockade. *European Journal of Neuroscience* 11:4309–4319.

Basso, A., Taborelli, A., and Vignolo, L. A. 1978. Dissociated disorders of speaking and writing aplasia. *Journal of Neurology, Neurosurgery, and Psychiatry* 41:556–563.

Beaton, A., and Mariën, P. 2010. Language, cognition and the cerebellum: Grappling with an enigma. *Cortex* 46:811–820.

Behrendt, J. E. 1984. Alzheimer's disease and its effects on handwriting. *Journal of Forensic Science* 29:87–91.

Bennet, D. A., Beckett, L. A., Murray, A. M., Shannon, K. M., Goetz, C. G., Pilgrim, D. M., and Evans, D. A. 1996. Prevalence of parkinsonian signs and associated mortality in a community population of older people. *New England Journal of Medicine* 334:71–76.

Berg, D., Becker, G., Zeiler, B., Tucha, O., Hofmann, E., Preier, M., Benz, P., et al. 1999. Vulnerability of the nigrostriatal system as detected by transcranial ultrasound. *Neurology* 53:1026–1031.

Bermejo-Pareja, F., Benito-León, J., Vega, S., Medrano, M. J., and Román, G. C. 2008. Incidence and subtypes of dementia in three elderly populations of central Spain. *Journal for Neurological Science* 264:63–72.

Bernstein, N. A. 1947. *On the structure of movements.* Moscow: State Medical Publishing House.

Bernstein, N. 1967. The coordination and regulation of movements. Oxford: Pergamon.

Betarbet, R., Sherer, T. B., and Greenamyre, J. T. 2002. Animal models of Parkinson's disease. *BioEssays* 24:308–318.

Bhidayasiri, R. 2005. Differential diagnosis of common tremor syndromes. *Postgraduate Medical Journal* 81:756–762.

Bloesch, E. K., and Abrams, R. A. 2010. Visuomotor binding in older adults. *Brain and Cognition* 74:239–243.

Boisseau, M., Chamberland, G., and Gauthier, S. 1987. Handwriting analysis of several extrapyramidal disorders. *Canadian Society of Forensic Science Journal* 20:139–146.

Bower, J. H., Maraganore, D. M., McDonnell, S. D. K., and Rocca, W. A. 1999. Incidence and distribution of parkinsonism in Olmsted County, Minnesota, 1976–1990. *Neurology* 52:1214–1220.

Braak, H., Ghebremedhin, E., Rub, U., Bratzke, H., and del Tredici, K. 2004. Stages in the development of Parkinson's disease-related pathology. *Cell Tissue Research* 318:121–124.

Brody, H. 1955. Organization of the cerebral cortex III. A study of aging in the human cerebral cortex. *Journal of Comparative Neurology* 102:511.

Brooks, D. J., Piccini, P., Turjanski, N., and Samuel, M. 2000. Neuroimaging of dyskinesia. *Annals of Neurology* 47 (Suppl 1): S154–158.

Brooks, V. B. 1986. *The neural basis of motor control.* New York: Oxford University Press.

Broussolle, E., Krack, P., Thobois, S., Xie-Brustolin, J., Pollak, P., and Goetz, C. G. 2007. Contribution of Jules Froment to the study of parkinsonian rigidity. *Movement Disorders* 22:909–914.

Brun, B., and Reisby, N. 1971. Handwriting changes following meprobamate and alcohol. *Quarterly Journal of Studies on Alcohol* 32:1070–1082.

Bryan, W. L. 1892. On the development of voluntary motor ability. *American Journal of Psychology* 5:125.

Bugiani, O., Salvarin, I. S., Perdelli, F., Mancardi, G. L., and Leonardi, A. 1978. Nerve cell loss with aging in the putamen. *European Neurology* 17:286–291.

Bullock, D., Grossberg, S., and Mannes, C. 1993. A neural network model for cursive script production. *Biological Cybernetics* 70:15–28.

Buquet, A. and Rudler M. 1987. Handwriting and exogenous intoxication. *International Criminal Police Review* 408: 9–20.

Burne, J. A., Hays, M. W., Fung, V. S. C., Yiannikas, C., and Boljevac, D. 2002. The contribution of tremor studies to diagnosis of parkinsonian and essential tremor: A statistical evaluation. *Journal of Clinical Neuroscience* 9:237–242.

Byrnes, M. L., Mastaglia, F. L., Walters, S. E., Archer, S. A., and Thickbroom, G. W. 2005. Primary writing tremor: Motor cortex reorganization and disinhibition. *Journal of Clinical Neuroscience* 12:102–104.

Caligiuri, M. P., and Buitenhuys, C. 2005. Do preclinical findings of methamphetamine-induced motor abnormalities translate to an observable clinical phenotype? *Neuropsychopharmacology* 30:2125–2134.

Caligiuri, M. P., and Galasko, D. R. 1992. Quantifying drug-induced changes in parkinsonian rigidity using an instrumental measure of activated stiffness. *Clinical Neuropharmacology* 15:1–12.

Caligiuri, M. P., Jeste, D. V., and Lacro, J. P. 2000. Antipsychotic-induced movement disorders in the elderly: Epidemiology and treatment recommendations. *Drugs and Aging* 17:363–384.

Caligiuri, M. P., and Lohr, J. B. 1993. Worsening of postural tremor in patients with levodopa-induced dyskinesia: A quantitative analysis. *Clinical Neuropharmacology* 16:244–250.

———. 1994. A disturbance in the control of muscle force in neuroleptic-naïve schizophrenic patients. *Biological Psychiatry* 15:104–111.

Caligiuri, M. P., Lohr, J. B., and Jeste, D. V. 1993. Parkinsonism in neuroleptic-naïve schizophrenic patients. *American Journal of Psychiatry* 150:1343–1348.

Caligiuri, M. P., Lohr, J. B., and Ruck, R. K. 1998. Scaling of movement velocity: A measure of neuromotor retardation in individuals with psychopathology. *Psychophysiology* 35:431–437.

Caligiuri, M. P., Rockwell, E., and Jeste, D. V. 1998. Extrapyramidal side effects in patients with Alzheimer's disease treated with low-dose neuroleptic medication. *American Journal Geriatric Psychiatry* 6:75–82.

Caligiuri, M. P., Teulings, H. L., Dean, C. E., Niculescu, A. B., and Lohr, J. B. 2009. Handwriting movement analyses for monitoring drug-induced motor side effects in schizophrenia patients treated with risperidone. *Human Movement Science* 28:633–642.

———. 2010. Handwriting movement kinematics for quantifying EPS in patients treated with atypical antipsychotics. *Psychiatry Research* 177:77–83.

Caligiuri, M. P., Teulings, H. L., Filoteo, J. V., Song, D., and Lohr, J. B. 2006. Quantitative measurement of handwriting in the assessment of drug-induced parkinsonism. *Human Movement Science* 25:510–22.

Calzetti, S., Baratti, M., Gresty, M., and Findley, L. 1987. Frequency/amplitude characteristics of postural tremor of the hands in a population of patients with bilateral essential tremor: Implications for the classification and mechanism of essential tremor. *Journal of Neurology, Neurosurgery, Psychiatry* 50:561–567.

Canales, J. J., Capper-Loup, C., Hu, D., Choe, E. S., Upadhyay, U., and Graybiel, A. M. 2002. Shifts in striatal responsivity evoked by chronic stimulation of dopamine and glutamate systems. *Brain* 125:2353–2363.

Carlsson, A. 1981. Aging and brain neurotransmitters. In *Funkitionsstorurgen des Gehirns im Alter,* ed. D. Platt, 67–81. Stuttgart: Schattauer.

Carlsson, A., and Winblad, B. 1976. The influence of age and time interval between death and autopsy on dopamine and 3-methoxytyramine levels in human basal ganglia. *Journal of Neural Transmissions* 38:271–276.

Carney, B. B. 1993. A new tremor in handwriting? Presented at the meeting of the American Society of Questioned Document Examiners, Ottawa, Ontario, Canada.

Chang, L., Ernst, T., Speck, O., Patel, H., DeSilva, M., Leonido-Yee, M., and Miller, E. N. 2002. Perfusion MRI and computerized cognitive test abnormalities in abstinent methamphetamine users. *Psychiatry Research* 114:65–79.

Chapman, D. E., Hanson, G. E., Kesner, R. P., and Keefe, K. A. 2001. Long-term changes in basal ganglia function after a neurotoxic regimen of methamphetamine. *Journal of Pharmacology and Experimental Therapeutics* 296:520–527.

Christou, E. A. 2011. Aging and variability of voluntary contractions. *Exercise and Sport Science Reviews* 39:77–84.

Cole, K. J. 1991. Grasp force control in older adults. *Journal of Motor Behavior* 23:251–258.

———. 2006. Age-related directional bias of fingertip force. *Experimental Brain Research* 175:285–291.

Cole, K. J., and Beck, C. L. 1994. The stability of precision grip force in older adults. *Journal of Motor Behavior* 26:171–177.

Cole, K. J., Rotella, D. L., and Harper, J. G. 1999. Mechanisms for age-related changes of fingertip forces during precision gripping and lifting in adults. *Journal of Neuroscience* 19:3238–3247.

Conneally, P. M. 1984. Huntington disease: Genetics and epidemiology. *American Journal of Human Genetics,* 36:506–526.

Contreras-Vidal, J. L., Teulings, H. L., and Stelmach, G. E. 1998. Elderly subjects are impaired in spatial coordination in fine motor control. *Acta Psychologica (Amsterdam)* 100:25–35.

Contreras-Vidal, J. L., Teulings, H. L., Stelmach, G. E., and Adler, C. H. 2002. Adaptation to changes in vertical display gain during handwriting in Parkinson's disease patients, elderly and young controls. *Parkinsonism and Related Disorders* 9:77–84.

Conway, J. 1959. *Evidential documents.* Springfield, IL: Charles C. Thomas.

Cordato, N. J., Halliday, G. M., Caine, D., and Morris, J. G. 2006. Comparison of motor, cognitive, and behavioral features in progressive supranuclear palsy and Parkinson's disease. *Movement Disorders* 21:632–638.

Cordato, N. J., Halliday, G. M., McCann, H., Davies, L., Williamson, P., Fulham, M., and Morris, J. G. L. 2001. Corticobasal syndrome with tau pathology. *Movement Disorders* 16:656–667.

Crary, M. A., and Heilman, K. M. 1988. Letter imagery deficits in a case of pure apraxic agraphia. *Brain Language* 34:147–156.

Croisile, B., Carmoi, T., Adeleine, P., and Trillet, M. 1995. Spelling in Alzheimer's disease. *Behavioral Neurology* 8:135–143.

Cruse, H. 1986. Constraints for joint angle control of the human arm. *Biological Cybernetics* 54:125–132.

Cummings, J. L. 1993. Frontal-subcortical circuits and human behavior. *Archives of Neurology* 50:873–880.

Darling, W., Cooke, J. D., and Brown, S. H. 1989. Control of simple arm movements in elderly humans. *Neurobiology of Aging* 10:149–157.

Davis, P. C., Mirra, S. S., and Alazraki, N. 2005. The brain in older persons with and without dementia: Findings on MR, PET, and SPECT images. *American Journal of Radiology* 162:1267–1287.

DeKeyser, J., Ebiner, G., and Vanquelin, G. 1990. Age-related changes in the human nigrostriatal dopamine system. *Annals of Neurology* 27:157–161.

DeLong, M. R. 1990. Primate models of movement disorders of basal ganglia origin. *Trends in Neurosciences* 13:281–285.

DeLong, M. R., and Wichmann, T. 2007. Circuits and circuit disorders of the basal ganglia. *Archives of Neurology* 64:20–24.

De Stefano, C., Marcelli, A., and Rendina, M. 2009. Disguising writer identification: An experimental study. *Proceedings of the International Graphonomics Society Conference,* Dijon, France.

Dewhurst, T., Found, B., and Rogers, D. 2008. Are expert penmen better than lay people at producing simulations of a model signature? *Forensic Science International* 180:50–53.

Di Carlo, A., Baldereschi, M., Amaducci, L., Lepore, V., Bracco, L., Maggi, S., Bonaiuto, S., et al. 2002. Incidence of dementia, Alzheimer's disease, and vascular dementia in Italy. The ILSA Study. *Journal of American Geriatric Society,* 50:41–48. Dickson, D. W. 2001. Alpha-synuclein and the Lewy body disorders. *Current Opinion in Neurology* 14:423–432.

Diermayr, G., McIsaac, T. L., and Gordon, A. M. 2011. Finger force coordination underlying object manipulation in the elderly—A mini review. *Gerontology* 57:217–227.

Diggles-Buckles, V. 1993. Age-related slowing. In *Sensorimotor impairment in the elderly,* ed. G. E. Stelmach and V. Homberg, 73–87. Norwell, MA: Kluwer Academic.

Di Monte, D. A., Lavasani, M., and Manning-Bog, A. B. 2002. Environmental factors in Parkinson's disease. *Neurotoxicology* 23:487–502.

Dixon, R. A., Kurzman, D., and Friesen, I. C. 1993. Handwriting performance in younger and older adults: Age, familiarity, and practice effects. *Psychology and Aging* 8:360–370.

Djioua M., and Plamondon, R. 2009. Studying the variability of handwriting patterns using the kinematic theory. *Human Movement Science* 28:588–601.

———. 2010. The limit profile of a rapid movement velocity. *Human Movement Science* 29:48–61.

Dostrovsky, J. O., Hutchison, W. D., and Lozano, A. M. 2002. The globus pallidus, deep brain stimulation, and Parkinson's disease. *Neuroscientist* 8:284–290.

Dounskaia, N., Van Gemmert, A. W. A., Leis, B. C., and Stelmach, G. E. 2009. Biased wrist and finger coordination in parkinsonian patients during performance of graphical tasks. *Neuropsychologia* 47:2504–2514.

D'Souza, D. C., Ranganathan, M., Braley, G., Gueorguieva, R., Zimolo, Z., Cooper, T., Perry, E., and Krystal, J. 2008. Blunted psychotomimetic and amnestic effects of delta-9-tetrahydrocannabinol in frequent users of cannabis. *Neuropsychopharmacology* 33:2505–2516.

Du Pasquier, R. A., Blanc, Y., Sinnreich, M., Landis, T., Burkhard, P., and Vingerhoets, F. J. 2003. The effects of aging on postural stability: A cross-sectional and longitudinal study. *Neurophysiologie Clinique* 33:213–218.

Durina, M. 2005. Disguised signatures: Random or repetitious? *Journal of American Society of Questioned Document Examiners* 8:9–16.

Eisen, A., Entezari-Taher, M., and Stewart, H. 1996. Cortical projections to spinal motoneuron changes with aging and amyotropic lateral sclerosis. *Neurology* 46:1394–1404.

Elble, R. J. 2002. Essential tremor is a monosymptomatic disorder. *Movement Disorders* 17:633–637.

———. 2003. Characteristics of physiologic tremor in young and elderly adults. *Clinical Neurophysiology* 114:624–635.

Elble, R. J., and Koller, W. C., 1990. *Tremor.* Baltimore, MD: The Johns Hopkins University Press.

Elble, R. J., Thomas, S. S., Higgens, C., and Colliver, J. 1991. Stride-dependent changes in gait of older people. *Journal of Neurology* 238:1–5.

Ellis, A. W. 1982. Spelling and writing (and reading and speaking). In *Normality and pathology in cognitive functions,* ed A. W. Ellis, 113–146. London: Academic Press.

Engelbrecht, S. E. 1997. Minimum-torque posture control. Doctoral dissertation. University of Massachusetts, Amherst. (University Microfilms No. 9721446).

———. 2001. Minimum principles in motor control. *Journal of Mathematical Psychology* 45:497–542.

Enoka, R. M., Christou, E. A., Hunter, S. K., Kornatz, K. W., Semmler, J. G., Taylor, A. M., and Tracy, B. L. 2003. Mechanisms that contribute to differences in motor performance between young and old adults. *Journal of Electromyography and Kinesiology* 13:1–12.

Era, P., Heikkinen, E., Gause-Nilsson, I., and Schroll, M. 2002. Postural balance in elderly people: Changes over a five-year follow-up and its predictive value for survival. *Aging Clinical and Experimental Research* 14:37–46.

Estabrooks, C. 2000. Measuring relative pen pressure to authenticate signatures. *Journal of American Society of Questioned Document Examiners* 3:57–65.

Evarts, E. V., and Tanji, J. 1974. Gating of motor cortex reflexes by prior instruction. *Brain Research* 71:479–494.

Exner, S. 1881. *Untersuchungen über Die Lokalisation Der Funktionen in Der Grosshirnrinde des Menschen.* Wein, Germany: Wilhelm Braunmüller.

Expert Consensus Panel 2003. The expert consensus guideline series: Optimizing pharmacologic treatment of psychiatric disorders. *Journal of Clinical Psychiatry* 64 (Suppl 12): 2–97.

Eyzaguirre, C., and Fidone, S. J. 1975. *Physiology of the nervous system,* 2nd ed. Chicago: Year Book Medical Publishers, Inc.

Fahn, S., and Elton, R. L. 1987. Unified Parkinson's disease rating scale. In *Recent developments in Parkinson's disease,* ed. S. Fahn, C. D. Marsden, D. Calne, and M. Goldstein, 153–164. Florham Park, NJ: Macmillan Health Care Information.

Findley, L. J., Gresty, M. A., and Halmagyi, G. M. 1981. Tremor, the cogwheel phenomenon and myoclonus in Parkinson's disease. *Journal of Neurology, Neurosurgery, and Psychiatry* 44:534–546.

Findley, L. J., and Koller, W. C. 1987. Essential tremor: A review. *Neurology* 37:1194–1197.

Finger, S. 1994. *Origins of neuroscience: A history of explorations into brain function.* New York: Oxford University Press.

Fitts, P. M. 1954. The information capacity of the human motor system in controlling the amplitude of movement. *Journal of Experimental Psychology* 47:381–391.

Fitzgerald, P. B., Williams, S., and Daskalakis, Z. L. 2009. A transcranial magnetic stimulation study of the effects of cannabis use on motor cortical inhibition and excitability. *Neuropsychopharmacology* 34:2368–2375.

Flash, T., and Hogan, N. 1985. The coordination of arm movements: An experimentally confirmed mathematical model. *Journal of Neuroscience* 5:1688–1703.

Flash, T., Inzelberg, R., Schechtman, E., and Korczyn, A. D. 1992. Kinematic analysis of upper limb trajectories in Parkinson's disease. *Experimental Neurology* 118:215–226.

Foley, R. G., and Miller, A. L. 1979. The effects of marijuana and alcohol usage on handwriting. *Forensic Science International* 14:159–164.

Forno, L. S. 1996. Neuropathology of Parkinson's disease. *Journal of Neuropathology & Experimental Neurology* 55:259–272.

Found, B., and Rogers, D. 2008. The probative character of forensic handwriting examiners' identification and elimination opinions on questioned signatures. *Forensic Science International* 178:54–60.

Found, B., Rogers, D., Rowe, V., and Dick, D. 1998. Statistical modeling of experts' perceptions of the ease of signature simulation. *Journal of Forensic Document Examination* 11:73–79.

Found, B., Sita, J., and Rogers, D. 1999. The development of a program for characterizing forensic handwriting examiners' expertise: Signature examination pilot study. *Journal of Forensic Document Examination* 12:69–79.

Found, B. and Rogers, D. 1996. The forensic investigation of signature complexity. In. Simmer, M., Leedham, G., and Thomassen, A. eds. *Handwriting and Drawing Research: Basic and Applied Issues*. Amsterdam, IOS Press 483–492.

Found, B. and Rogers, D. 1998. A consideration of the theoretical basis of forensic handwriting examination: The application of 'Complexity Theory" to understanding the basis of handwriting identification. *International Journal of Forensic Document Exam* 4: 109–118.

Found, B. and Rogers, D. 1999. The development of a program for characterizing forensic handwriting examiners' expertise: signature examination pilot study. *International Journal of Forensic Document Exam* 12: 69–80.

Fozard, J. L., Vercryssen, M., Reynolds, S. L., Hancock, P. A., and Quilter, R. E. 1994. Age differences and changes in reaction time: The Baltimore longitudinal study on aging. *Journal of Gerontology* 49:179–189.

Franke, K. 2009. Analysis of authentic signatures and forgeries. *Computational forensics. Third International Workshop Proceedings, IWCF 2009*, ed. Z. Geradts, K. Franke, and C. Veenman. The Hague, The Netherlands, August 13–14.

Frye, G. D., and Breese G. R. 1982. GABAergic modulation of ethanol-induced motor impairment. *Journal of Pharmacology and Experimental Therapeutics* 233:750–765.

Galbraith, N. G. 1986. Alcohol: Its effect on handwriting. *Journal of Forensic Science* 31:580–588.

Galganski, M. E., Fuglevand, A. J., and Enoka, R. M. 1993. Reduced control of motor output in a human hand muscle of elderly subjects during submaximal contractions. *Journal of Neurophysiology* 69:2108–2115.

Galton, F. 1899. Exhibition of instruments (1) for testing perception of differences of tint and (2) for determining reaction time. *Journal of Anthropology Institute* 19:27–29.

Geller, E. S., Clarke, S. W., and Kalsher, M. J. 1991. Knowing when to say when: a simple assessment of alcohol impairment. *Journal of Applied Behavioral Analysis* 24: 65–72.

Gerber, P. E., and Lynd, L. D. 1998. Selective serotonin reuptake inhibitor-induced movement disorders. *Annals of Pharmacotherapy* 32:692–698.

Gerken, A., Wetzel, H., and Benkert ,O. 1991. Extrapyramidal symptoms and their relationship to clinical efficacy under perphenazine treatment. A controlled prospective handwriting-test study in 22 acutely ill schizophrenic patients. *Pharmacopsychiatry* 24: 132–137.

Geser, F., Wenning, G. K., Poewe, W., and McKeith, I. 2005. How to diagnose dementia with Lewy bodies: State of the art. *Movement Disorders* Suppl 1, S11–S20.

Gessell, H. J. E. 1961. Drugs and questioned document problems. *Journal of Forensic Sciences* 6: 76–87.

Ghilardi, M.F., Alberoni, M., Rossi, M., Franceschi, M., Mariani, C., and Fazio, F. 2000. Visual feedback has differential effects on reaching movements in Parkinson's and Alzheimer's disease. *Brain Research* 876:112–23

Gill, J., Allum, J. H., Carpenter, M. G., Held-Ziolkowska, M., Adkin, A. L., Honegger, F., and Pierchala, K. 2001. Trunk sway measures of postural stability during clinical balance tests: Effects of age. *Journal of Gerontology* 56:M438–M447.

Gillig, P. M., and Sanders, R. D. 2010. Psychiatry, neurology, and the role of the cerebellum. *Psychiatry* 7:38–43.

Gilmour, C., and Bradford, J. 1987. The effect of medication on handwriting. *Canadian Society of Forensic Science Journal* 20:119–138.

Glencross, D. J. 1977. Control of skilled movement. *Psychological Bulletin* 84:14–29.

Godfrey, A., Conway, R., Meagher, D. and O'Laighin, G. 2008. Direct measurement of human movement by accelerometry. *Medical Engineering and Physics,* 30:1364–1386.

Gross, L. J. 1975. Drug-induced handwriting changes: An empirical review. *Texas Reports on Biology and Medicine* 33:371–390.

Grossberg, S., and Paine, R. 2000. A neural model of corticocerebellar interactions during attentive imitation and predictive learning of sequential movements. *Neural Networks* 13:999–1046.

Guest, R., Fairhurst, M., and Linnell, T. 2009. Towards an inferred data accuracy assessment of forensic document examination methodologies for signatures. *Proceedings of the International Graphonomics Society.* Dijon, France.

Guilarte, T. R. 2001. Is methamphetamine abuse a risk factor in parkinsonism? *Neurotoxicology* 22:725–731.

Haase, H. J. 1961. Extrapyramidal modification of fine movements: a "conditio sine qua non" of the fundamental therapeutic action of neuroleptic drugs. *Review of Canadian Biology* 20:425–49.

Haase, H. J. 1978. The purely neuroleptic effect and its relation to the "neuroleptic threshold." *Acta Psychiatrica Belgica* 78:19–36.

Haase, H. J., and Janssen, P. A. J. 1965. *The action of neuroleptic drugs: A psychiatric, neurologic, and pharmacologic investigation.* Chicago: Year Book Medical Publishers.

Hai, C., Yu-Ping, W., Hua, W., and Ying, S. 2010. Advances in primary writing tremor. *Parkinsonism and Related Disorders* 16:561–565.

Hall, W., and Degenhardt, L. 2009. Adverse health effects of non-medical cannabis use. *Lancet* 374:1383–1391.

Hallett, M. 1998. Overview of human tremor physiology. *Movement Disorders* 13 (Suppl 3): 43–48.

Hallet, M., and Koshbin, S. 1980. A physiological mechanism of bradykinesia. *Brain* 103:301–314.

Han, Z., Kuyatt, B. L., Kochman, K. A., DeSouza, E. B., and Roth, G. S. 1989. Effect of aging on concentrations of D2-receptor-containing neurons in the rat striatum. *Brain Research* 498:299–307.

Hanchar, H. J., Dodson, P. D., Olsen, R. W., Otis, T. S., and Wallner, M. 2005. Alcohol-induced motor impairment caused by increased extrasynaptic GABA A receptor activity. *Nature Neuroscience* 8:339–345.

Hansen, L., and Galasko, D. R. 1992. Lewy body disease. *Current Opinions in Neurology & Neurosurgery* 5:889–894.

Hanyu, H., Sato, T., Hirao, K., Kanetaka, H., Sakurai, H., and Iwamato, T. 2009. Differences in clinical course between dementia with Lewy bodies and Alzheimer's disease. *European Journal of Neurology* 16:212–217.

Harralson, H. H., Teulings, H. L., and Farley, B. G. 2009. Handwriting variability in movement disorder patients with fatigue. *Advances in Graphonomics: Proceedings of IGC,* 103–107.

Harrison, W. 1958. *Suspect documents: Their scientific examination.* New York: Praeger.

Haycock, J. W., Becker, L., Ang, L., Furukawa, Y., Hornykiewicz, O., and Kish, S. J. 2003. Marked disparity between age-related changes in dopamine and other presynaptic dopaminergic markers in human striatum. *Journal of Neurochemistry* 87:574–585.

Hebert, L. E., Scherr, P. A., Bienias, J. L., Bennett, D. A., and Evans, D. A. 2003. Alzheimer disease in the US population: Prevalence estimates using the 2000 census. *Archives of Neurology* 60:1119–1122.Hegerl, U., Mergl, R., Henkel, V., Pogarell, O., Muller-Siecheneder, F., Frodl, T., and Juckel, G. 2005. Differential effects of reboxetine and citalopram on hand-motor function in patients suffering from major depression. *Psychopharmacology* 178:58–66.

Hennebert, J., Loeffel, R., Humm, A., and Ingold, R. 2007. A new forgery scenario based on regaining dynamics of signature. In *Lecture notes in computer science,* vol. 4642/IBC 2007, ed. S. W. Lee and S. Z. Li, 366–375. Berlin: Springer–Verlag.

Henry, B., Duty, S., Fox, S. H., Crossmand, A. R., and Brotchie, M. 2003. Increased striatal pre-proenkephalin B expression is associated with dyskinesia in Parkinson's disease. *Experimental Neurology* 183:458–468.

Henry, J. M., and Roth, G. S. 1984. Effect of aging on recovery of striatal dopamine receptors following *N*-ethoxycarbonyl-2-ethoxy-1,2-dihydroquino-line (EEDQ) blockade. *Life Science* 35:899–904.

Herkt, A. 1986. Signature disguise or signature forgery. *Journal of Forensic Science Society* 26:257–266.

Hilton, O. 1961. Contrasting effects of forgery and genuineness. Paper presented at the Annual Conference of the American Society of Questioned Document Examiners. Los Angeles, CA.

———. 1962. Handwriting and the mentally ill. *Journal of Forensic Science* 7:131–139.

———. 1969. A study of the influence of alcohol on handwriting. *Journal of Forensic Science* 14:309–316.

———. 1977. Influence of age and illness on handwriting: Identification problems. *Forensic Science* 9:161–172.

Hinde, R. A. 1969. Control of movement patterns in animals. *Quarterly Journal of Experimental Psychology* 21:105–126.

Hindle, J. V. 2010. Ageing, neurodegeneration and Parkinson's disease. *Age and Ageing* 39:156–161.

Hogan, N. 1984. An organizing principle for a class of voluntary movements. *Journal of Neuroscience* 4:2745–2754.

Hollerbach, J. M. 1981. An oscillation theory of handwriting. *Biological Cybernetics* 39:139–156.

Hollerbach, J. M., and Flash, T. 1982. Dynamic interactions between limb segments during planar arm movement. *Biological Cybernetics* 44:67–77.

Horak, F. B., and Anderson, M. E. 1984. Influence of globus pallidus on arm movements in monkeys. I. Effects of kainic acid-induced lesions. *Journal Neurophysiology* 52:290–304.

Houk, J. C., and Wise, S. P. 1995. Distributed modular architectures linking basal ganglia, cerebellum, and cerebral cortex: Their role in planning and controlling action. *Cerebral Cortex* 2:95–110.

Huber, R. A., and Headrick, A. M. 1999. *Handwriting identification: Facts and fundamentals.* Boca Raton, FL: CRC Press LLC.

Hughes, J. C., Graham, N., Patterson, K., and Hodgens, J. 1997. Dysgraphia in mild dementia of Alzheimer's type. *Neuropsychologia* 4:533–545.

Hulstijn, W., and Van Galen, G. P. 1988. Levels of motor programming in writing familiar and unfamiliar symbols. In *Cognition and action in skilled behavior,* ed. A. M. Colley and J. R. Beech, 65–85. Amsterdam: Elsevier Science Publishers.

Hunault, C. C., Mensinga, T. T., Bocker, K. B., Schipper, C. M., Kruidenier, M., Leenders, M. E., de Vries, I., and Meulenbelt, J. 2009. Cognitive and psychomotor effects in males after smoking a combination of tobacco and cannabis containing up to 69 mg delta-9-tetrahydrocannabinol (THC). *Psychopharmacology (Berl)* 204:85–94.

Ibanez, V., Sadato, N., Karp, B., Deiber, M. P., and Hallett, M. 1999. Deficient activation of the motor cortical network in patients with writer's cramp. *Neurology* 53:96–105.

Ishida, Y., Kozaki, T., Isomura, Y., Ito, S., and Isobe, K. 2009. Age-dependent changes in dopaminergic neuron firing patterns in substantia nigra pars compacta. *Journal of Neural Transmission* (Suppl) 73:129–133.

Iwasaki, Y., Ikeda, K., Shindo, T., Suga, Y., Ishikawa, I., Hara, M., and Shibamoto, A. 1999. Micrographia in Huntington's disease. *Journal of the Neurological Sciences* 162:106–107.

Jankovic, J. 2002. Essential tremor: A heterogeneous disorder. *Movement Disorders* 17:638–644.

———. 2008. Parkinson's disease: Clinical features and diagnosis. *Journal of Neurology, Neurosurgery, and Psychiatry* 79:368–376.

Jellinger, K., and Mizuno, Y. 2003. Parkinson's disease. In *Neurodegeneration: The molecular pathology of dementia and movement disorders,* ed. D. Dickson, 159–187. Basel, Switzerland: ISN Neuropathology Press.

Jeste, D. V., Caligiuri, M. P., Paulsen, J. S., Heaton, R. K., Lacro, J. P., Harris, M. J., Bailey, A., Fell, R. L., and McAdams, L. A. 1995. Risk of tardive dyskinesia in older patients: A longitudinal study of 266 patients. *Archives of General Psychiatry* 52:756–765.

Jimenez-Jimenez, F. J., Cabrera-Valdivia, F., Orti-Pareja, M., Gasalla, T., Tallon-Barranco, A., and Zurdo, M. 1998. Bilateral primary writing tremor. *European Journal of Neurology* 5:613–614.

Kachi, T., Rothwell, J. C., Cowan, J. M., and Marsden, C. D. 1985. Writing tremor: Its relationship to benign essential tremor. *Journal of Neurology, Neurosurgery, and Psychiatry* 48:545–550.

Kalechstein, A. D., Newton, T. F., and Green, M. 2003. Methamphetamine dependence is associated with neurocognitive impairment in the initial phases of abstinence. *Journal of Neuropsychiatry and Clinical Neuroscience* 15:215–220.

Kam, M., Fielding, G., and Conn, R. 1998. Effects of monetary incentives on performance of nonprofessionals in document examination proficiency tests. *Journal of Forensic Science* 43:1000–1004.

Kam, M., Gummadidala, K., Fielding, G., and Conn, R. 2001. Signature authentication by forensic document examiners. *Journal of Forensic Science* 46:884–888.

Kamen, G., and Roy, A. 2000. Motor unit synchronization in young and elderly adults. *European Journal of Physiology and Occupational Physiology* 81:403–410.

Kaye, J. A., Oken, B. S., Howieson, D. B., Howieson, J., Holm, L. A., and Dennison, K. 1994. Neurologic evaluation of the optimally healthy oldest old. *Archives of Neurology* 51:1205–1211.

Keele, S. W. 1968. Movement control in skilled motor performance. *Psychological Bulletin* 70:387–403.

———. 1981. Behavioral analysis of movement. In *Handbook of physiology: Vol. 2. Motor control, part 2,* ed. V. B. Brooks, 1391–1414. Baltimore, MD: American Physiological Society.

Keele, S. W., and Summers, J. J. 1976. The structure of motor programs. In *Motor control: Issues and trends,* ed. G. E. Stelmach, 109–142. New York: Academic Press.

Keepers, G. A., and Casey, D. E. 1991. Use of neuroleptic-induced extrapyramidal symptoms to predict future vulnerability to side effects. *American Journal of Psychiatry* 148:85–89.

Kelsoe, J. A. S., Holt, K. G., Rubin, P., and Kugler, P. N. 1981. Patterns of human interlimb coordination emerge from the properties of nonlinear, limit-cycle, oscillatory processes: Theory and data. *Journal of Motor Behavior* 13:226–261.

Kim, J. H., and Byun, H. J. 2010. The relationship between akathisia and subjective tolerability in patients with schizophrenia. *International Journal of Neuroscience* 120:507–511.

Kim, J. S., Im, J. H., Kwon, S. U., Kang, J. H., and Lee, M. C. 1998. Micrographia after thalamomesencephalic infarction: Evidence of striatal dopaminergic hypofunction. *Neurology* 51:625–627.

King, M. B., Judge, J. O., and Wolfson, L. 1994. Functional base of support decreases with age. *Journal of Gerontology* 49:M258–263.

Kolb, B., Forgie, M., Gibb, R., Gorny, G., and Rowntree, S. 1998. Age, experience, and the changing brain. *Neuroscience Biobehavioral Review* 22:143–159.

Koller, W. C. 1984. The diagnosis of Parkinson's disease. *Archives of Internal Medicine* 144:2146–2147.

Kornhuber, H. H. 1971. Motor functions of cerebellum and basal ganglia: The cerebellocortical saccadic (ballistic) clock, the cerebellonuclear hold regulator, and the basal ganglia ramp (voluntary speed smooth movement) generator. *Kybernetik* 8:157–162.

Kuenstler, U., Juhnhold, U., Knapp, W. H., and Gertz, H. H. 1999. Positive correlation between reduction of handwriting area and D2 dopamine receptor occupancy during treatment with neuroleptic drugs. *Psychiatric Research* 90: 31–39.

Kuenstler, U., Hohdorf, K., Regenthan, R., Seese, A., Gertz, H.J. 2000. Diminution of hand writing area and D2-dopamine receptor occupancy blockade: Results from treatment with typical and atypical neuroleptics. Nervenarzt 71: 373–379.

Kugler, P. N. and Turvey, M. T. 1987. *Information, natural law and the self-assembly of rhythmic movement.* Hillsdale NJ: Erlbaum.

Lacquaniti, F., Terzuolo, C. A., and Viviani, P. 1983. The law relating kinematic and figural aspects of drawing movements. *Acta Psychologica* 54:115–130.

———. 1984. Globalmetric properties and preparatory processes in drawing movements. In *Preparatory states and processes,* ed. S. Kornblum and J. Requin, 357–370. Hillsdale, NJ: Lawrence Erlbaum Associates.

Laidlaw, D. H., Bilodeau, M., and Enoka, R. M. 2000. Steadiness is reduced and motor unit discharge is more variable in old adults. *Muscle and Nerve* 23:600–612.

Lange, K. W., Mecklinger, L., Walitza, S., Becker, G., Gerlach, M., Naumann, M., and Tucha, O. 2006. Brain dopamine and kinematics of graphomotor functions. *Human Movement Science* 25:492–509.

Langston, J. W., Ballard, P., Tetrud, J. W., and Irwin, I. 1983. Chronic parkinsonism in humans due to a product of meperidine-analog synthesis. *Science* 219:979–980.

Lashley, K. S. 1931. Mass action in cerebral function. *Science* 73(1888): 245–54.

Lashley, K. S. 1951. The problem of serial order in behavior. In *Cerebral mechanisms in behavior. The Hixon symposium,* ed. L. A. Jeffress, 112–136. New York: John Wiley & Sons.

Latash, M. L., Shim, J. K., Shinohara, M., and Zatsiorsky, V. M. 2006. Changes in finger coordination and hand function with age. In *Motor control and learning,* ed. M. L. Latash and F. Lestienne, 141–159. New York: Springer.

Latash, M. L., Shim, J. K., and Zatsiorsky, V. M. 2004. Is there a timing synergy during multi-finger production of quick force pulses? *Experimental Brain Research* 159:65–71.

Lee, C. R., and Tepper, J. M. 2009. Basal ganglia control of substantia nigra dopaminergic neurons. *Journal of Neural Transmission* (Suppl) 73:71–90.

Legge, D. A. W. 1964. Simple measures of handwriting as indices of drug effects. *Personality and Motor Skills* 18: 549–558.

Leo, R. J. 1996. Movement disorders associated with serotonin selective reuptake inhibitors. *Journal of Clinical Psychiatry* 57:449–454.

Leung, S. C., Cheng, Y. S., Fung, H. T., and Poon, N. L. 1993. Forgery 1—Simulation. *Journal of Forensic Science* 38:402–412.

Levine, D. N., Mani, R. B., and Calvario, R. 1988. Pure agraphia and Gerstmann's syndrome as a visuospatial-language dissociation: An experimental case study. *Brain Language* 35:172–196.

Lewitt, P. A. 1983. Micrographia is a focal sign of neurological disease. *Journal of Neurology, Neurosurgery, and Psychiatry* 46:1152–1157.

Lindberg, P., Ody, C., Feydy, A., and Maier, M. A. 2009. Precision in isometric precision grip force is reduced in middle-aged adults. *Experimental Brain Research* 193:213–224.

Lipsitz, L. A. 2004. Physiological complexity, aging, and the path to frailty. *Science of Aging Knowledge Environment* pe16.

Llinas, R. 2009. Inferior olive oscillations as the temporal basis for motricity and oscillatory reset as the basis for motor error correction. *Neuroscience* 162:797–804.

Longstaff, M. G., Mahant, P. R., Stacy, M. A., van Gemmert, A. W., Leis, B. C., and Stelmach, G. E. 2001. Continuously scaling a continuous movement: Parkinsonian patients choose a smaller scaling ration and produce more variable movements compared to elderly controls. In *Proceedings of the Tenth*

Biennial Conference of the International Graphonomics Society, Nijmegen, ed. R. G. L. Meulenbroeck and B. Steenbergen, 84–89. Nijmegen, The Netherlands: IGS.

Louis, E. D. 2001. Samuel Adams' tremor. *Neurology* 56:1201–1205.

Louis, E. D., Ford, B., and Barnes, L. F. 2000. Clinical subtypes of essential tremor. *Archives of Neurology* 57:1194–1198.

Louis, E. D., Ford, B., Pullman, S., and Baron, K. 1998. How normal is "normal"? Mild tremor in a multiethnic cohort of normal subjects. *Archives of Neurology* 55:222–227.

Louis, E. D., and Kavanaugh, P. 2005. John Adams' essential tremor. *Movement Disorders Society* 20:1537–1542.

Lozano, A. M., and Lang, A. E. 2001. Pallidotomy for Parkinson's disease. *Advances in Neurology* 86:413–420.

Lublin, F. D., and Reingold, S. C. 1996. Defining the clinical course of multiple sclerosis: Results of an international survey. National Multiple Sclerosis Society (USA) Advisory Committee on Clinical Trials of New Agents in Multiple Sclerosis. *Neurology* 46:907–911.

Lubrano, V., Roux, F. E., and Démonet, J. 2004. Writing-specific sites in frontal areas: A cortical stimulation study. *Journal of Neurosurgery* 101:787–798.

Lucchinetti, C. 2008. Pathological heterogeneity of idiopathic central nervous system demyelinating disorders. *Current Topics in Microbiology Immunology* 318:19–43.

Luzatti, C., Laiacona, M., Allamano, N., De Tanti, A., and Inzaghi, M. G. 1998. Writing disorders in Italian aphasic patients: A multiple single-case study of dysgraphia in a language with shallow orthography. *Brain* 121:1721–1734.

Luzatti, C., Laiacona, M., and Hagáis, D. 2003. Multiple patterns of writing disorders in dementia of the Alzheimer type and their evolution. *Neuropsychologia* 41:759–772.

Maarse, F., and Thomassen, A. 1983. Produced and perceived writing slant: Difference between up and down strokes. *Acta Psychologica* 54:131–147.

Magrassi, L., Bongetta, D., Bianchini, S., Berardesca, M., and Arienta, C. 2010. Central and peripheral components of writing critically depend on a defined area of the dominant superior parietal gyrus. *Brain Research* 1346:145–154.

Maki, B. E., Holliday, P. J., and Fernie, G. R. 1990. Aging and postural control: A comparison of spontaneous and induced-sway balance tests. *Journal of American Geriatric Society* 38:1–9.

Manzanares, J., Urigüen, L., Rubio, G., and Palomo, T. 2004. Role of endocannabinoid system in mental diseases. *Neurotoxicity Research* 6:213–224.

Margolin, D. I. 1984. The neuropsychology of writing and spelling: Semantic, phonological, motor, and spelling processes. *Quarterly Journal of Experimental Psychology* 36A:459–489.

Margolin, D. I., and Wing, A. M. 1983. Agraphia and micrographia: Clinical manifestations of motor programming and performance disorders. *Acta Psychologica* 54:263–283.

Marsden, C. D., Merton, P. A., Morten, H. B., Hallett, M., Adam, J., and Rushton, D. N. 1977. Disorders of movement in cerebellar disease in man. In *Physiologic aspects of clinical neurology*, ed. F. C. Rose, 179–199. London: Blackwell Press.

Marsden, C. D., Obeso, J., and Rothwell, J. C. 1983. Benign essential tremor is not a single entity. In *Current concepts in Parkinson's disease*, ed. M. D. Yahr, 31–46. Amsterdam: Excerpta Medica.

Marshall, J. 1961. The effect of ageing upon physiological tremor. *Journal of Neurology, Neurosurgery & Psychiatry* 24:14–17.

Matsuda, L. A., Lolait, S. J., Brownstein, M. J., Young, A. C., and Bonner, T. I. 1990. Structure of a cannabinoid receptor and functional expression of the cloned cDNA. *Nature* 346:561–564.

McGeer, E. G. 1978. Aging and neurotransmitter function in the human brain. In *Alzheimer's disease: Senile dementia and related disorders*. Aging series, vol. 7, ed. R. Katzman, R. D. Terry, and K. L. Bick, 427. New York: Raven Press.

McGeer, P. L., and McGeer, E. G. 1976. Enzymes associated with the metabolism of catecholamines, acetylcholine, and GABA in human controls and patients with Parkinson's disease and Huntington's chorea. *Journal of Neurotransmission* 26:65–76.

———. 1978. Aging and neurotransmitter systems. *Advances in Experimental Medicine and Biology* 113:41–57.

McGeer, P. L., McGeer, E. G., and Suzuki, J. S. 1977. Aging and extrapyramidal function. *Archives of Neurology* 34:33–35.

McKeith, I., Fairbairn, A., Perry, R., Thompson, P., and Perry, E. 1992. Neuroleptic sensitivity in patients with senile dementia of Lewy body type. *British Medical Journal* 305:673–678.

McKeith, I. G., Galasko, D., Kosaka, K., Perry, E. K., Dickson, D. W., Hansen, L. A., Salmon, D. P., et al. 1996. Consensus guidelines for the clinical and pathologic diagnosis of dementia with Lewy bodies (DLB): Report of the consortium on DLB international workshop. *Neurology* 47:1113–1124.

McLennan, J. E., Nakano, K., Tyler, H. R., and Schwab, R. S. 1972. Micrographia in Parkinson's disease. *Journal of the Neurological Sciences* 15:141–152.

McNay, E. C., and Willingham, D. B. 1998. Deficit in learning of a motor skill requiring strategy, but not of perceptuomotor recalibration, with aging. *Learning and Memory* 4:411–420.

Meeks, M., and Kuklinski, T. 1990. Measurement of dynamic digitizer performance. *Computer Processing of Handwriting*, ed. R. Plamondon and C. G. Leedham, 89–110. Singapore: World Scientific Publishing.

Melamed, E., Lavy, S., Bentin, S., Cooper, G., and Rinot, Y. 1980. Reduction in regional cerebral blood flow during normal aging in men. *Stroke* 11:31–35.

Menon, V., and Desmond, J. E. 2001. Left superior parietal cortex involvement in writing: Integrating fMRI with lesion evidence. *Cognitive Brain Research* 12:337–340.

Mergl, R., Juckel, G., Rihl, J., Henkel, V., Karner, M., Tigges, P., Schroter, A., and Hegerl, U. 2004. Kinematic analysis of handwriting in depressed patients. *Acta Psychiatrica Scandinavica* 109:383–391.

Messinis, L., Kyprianidou, A., Malefaki, S., and Papathanasopoulos, P. 2006. Neuropsychological deficits in long-term frequent cannabis users. *Neurology* 66:737–739.

Meyer, D. E., Abrams, R. A., Kornblum, S., Wright, C. E., and Smith, J. E. K. 1988. Optimality in human motor performance: Ideal control of rapid aimed movements. *Psychology Review* 95:340–370.

Miall, R. C., and Haggard, P. 1995. The curvature of human arm movements in the absence of visual experience. *Experimental Brain Research* 103:421–428.

Michel, L. 1978. Disguised signatures. *Journal of Forensic Science Society* 18:25–29.

Mink, J. W. 2003. The basal ganglia and involuntary movements—Impaired inhibition of competing motor patterns. *Archives of Neurology* 60:1365–1368.

Mink, J. W., and Thatch, W. T. 1993. Basal ganglia intrinsic circuits and their role in behavior. *Current Opinion in Neurobiology* 3:950–957.

Modugno, N., Nakamara, Y., Bestmann, S., Curra, A., Berardelli, A., and Rothwell, J. 2002. Neurophysiological investigations in patients with primary writing tremor. *Movement Disorders* 17:1336–1340.

Mohammed, L. 1993. Signature disguise in Trinidad and Tobago. *Science and Justice* 1:21–24.

Mohammed, L., Found, B., Caligiuri, M., and Rogers, D. 2010. The dynamic character of disguise behavior for text-based, mixed, and stylized signatures. *Journal of Forensic Sciences* 56:S136–S141.

Moore, S. P., and Marteniuk, R. G. 1986. Kinematic and electromyographic changes that occur as a function of learning a time-constrained aiming task. *Journal of Motor Behavior* 18:397–426.

Morasso, P. 1981. Spatial control of arm movements. *Experimental Brain Research* 42:223–227.

Morgan, D. G., and Finch, C. E. 1988. Dopaminergic changes in the basal ganglia. A generalized phenomenon of aging in mammals. *Annals of New York Academy of Sciences* 515:145–160.

Morris, R. 2000. *Forensic handwriting identification*. San Diego: Academic Press.

Morrish, P.K., Sawle, G.V., Brooks, D.J. 1996. An [18F]dopa-PET and clinical study of the rate of progression in Parkinson's disease. *Brain* 119: 585–591.

Moszczynska, A., Fitzmaurice, P., Ang, L., Kalasinsky, K. S., Schmunk, G. A., Peretti, F. J., Aiken, S. S., Wickham, D. J., and Kish, S. J. 2004. Why is parkinsonism not a feature of human methamphetamine abusers? *Brain* 127:363–370.

Mozley, P. D., Acton, P. D., Barraclough, E. D., Plossl, K., Gur, R. C., Alavi, A., Mathur, A., Saffer, J., and Kunk, H. F. 1999. Effects of age on dopamine transporters in healthy humans. *Journal of Nuclear Medicine* 40:1812–1817.

Muehlberger, R. 1990. Identifying simulations: Practical considerations. *Journal of Forensic Science* 35:368–374.

Muller, F., and Stelmach, G. E. 1992. Prehension movements in Parkinson's disease. In *Tutorials in motor behavior II,* ed. G. E. Stelmach and J. Requin, 307–319. Amsterdam: Elsevier Science.

Murray, R., Neumann, M., Forman, M. S., Farmer, J., Massimo, L., Rice, A., Miller, B. L., et al. 2007. Cognitive and motor assessment in autopsy-proven corticobasal degeneration. *Neurology* 68:1274–1283.

Nahhas, R. W., Choh, A. C., Lee, M., Cameron, W. M., Duren, D. L., Siervogel, R. M., Sherwood, R. J., Towne, B., and Czerwinski, S. A. 2010. Bayesian longitudinal plateau model of adult grip strength. *American Journal of Human Biology* 22:648–656.

Nakamura, M., Hamamoto, M., Uchida, S., Nagayama, H., Amemiya, S., Okubo, S., and Tanaka, K. 2003. A case of micrographia after subcortical infarction: Possible involvement of frontal lobe function. *European Journal of Neurology* 10:593–596.

Nance, M. A., and Myers, R. H. 2001. Juvenile onset Huntington's disease—Clinical and research perspectives. *Mental Retardation and Developmental Disabilities Research Reviews* 7:153–157.

Newell, K. M., Vaillancourt, D. E., and Sosnoff, J. J. 2006. Aging, complexity, and motor performance: Healthy and disease states. In *Handbook of the psychology of aging,* ed. J. Birren and K. Schaie, 163–182. Amsterdam: Elsevier.

Nicosia, N. 2009. The cost of methamphetamine use: a national estimate. Drug Policy Research Center. Rand. http://www.rand.org/pubs/research_briefs/RB9438.html

NIDA (National Institute on Drug Abuse) 2009. The economic costs of drug abuse in the United States, 1992–1998 (www.nida.nih.gov/InfoFacts). Office of National Drug Control Policy. Washington, DC.

Ohno, T., Bando, M., Natura, H., Ishi, K., and Yamanouchi, H. 2000. Apraxia agraphia due to thalamic infarction. *Neurology* 27:2336–2339.

Ondo, W. G., Wang, A., Thomas, M., and Vuong, K. D. 2005. Evaluating factors that can influence spirography ratings in patients with essential tremor. *Parkinsonism and Related Disorders* 11:45–48.

Osborn, A. S. 1929. *Questioned documents,* 2nd ed. Montclair, NJ: Patterson Smith.

Otsuki, M., Soma, Y., Arai, T., Otsuka, A., and Tsuji, S. 1999. Pure apraxic agraphia with abnormal writing stroke sequences: Report of a Japanese patient with a left superior parietal hemorrhage. *Journal of Neurology, Neurosurgery, and Psychiatry* 66:233–237.

Owens, M. C. 1990. Expectations of writing associated with advanced age and specific medical disorders: A look at document examiner and medical opinions. Paper presented to the American Society of Questioned Documents Examiners. San Jose, CA.

Pahl, J. J., Mazziotta, J. C., Bartzokis, G., Cummings, J., Altshuler, L., Mintz, J., Marder, S. R., and Phelps, M. E. 1995. Positron-emission tomography in tardive dyskinesia. *Journal of Neuropsychology Clinical Neuroscience* 7:457–465.

Parrott, A. C., Buchanan, T., Scholey, A. B., Heffernan, T., Ling, J., and Rodgers, J. 2002. Ecstasy/MDMA attributed problems reported by novice, moderate, and heavy users. *Human Psychopharmacology* 17:309–312.

Paulson, G. 2004. Illnesses of the brain in John Quincy Adams. *Journal of the History of Neuroscience.* 13: 336–44.

Pearce, J. M. S. 2008. The micrographia of Sir Henry Head (1861–1940). *Journal of Neurology, Neurosurgery, and Psychiatry* 79:307–308.

Penfield, W., and Rasmussen, T. 1950. *The cerebral cortex of man: A clinical study of localization of function.* New York: MacMillan.

Perry, R. H., Irving, D., and Thomlinson, B. E. 1990. Lewy body prevalence in the aging brain: Relationship to neuropsychiatric disorders, Alzheimer-type pathology and catecholaminergic nuclei. *Journal of Neurological Science* 100:223–233.

Pfann, K. D., Buchman, A. S., Comella, C. L., and Corcos, D. M. 2001. Control of movement distance in Parkinson's disease. *Movement Disorders* 16:1048–1065.

Pfann, K. D., Penn, R. D., Shannon, K. M., and Corcos, D. M. 1998. Pallidotomy and bradykinesia: Implications for basal ganglia function. *Neurology* 51:796–803.

Phillips, J. G. 2008. Can the relationship between tangential velocity and radius of a curvature explain motor constancy? *Human Movement Science* 27:799–811.

Phillips, J. G., Bradshaw, J. L., Chiu, E., and Bradshaw, J. A. 1994. Characteristics of handwriting of patients with Huntington's disease. *Movement Disorders* 9:521–530.

Phillips, J. G., Chiu, E., Bradshaw, J. L., and Iansek, R. 1995. Impaired movement sequencing in patients with Huntington's disease: A kinematic analysis. *Neuropsychologia* 33:365–369.

Phillips, J. G., Ogeil, R. P., and Muller, F. 2009. Alcohol consumption and handwriting: A kinematic analysis. *Human Movement Science* 28:619–632.

Phillips, J. G., Stelmach, G. E., and Teasdale, N. 1991. What can indices of handwriting quality tell us about parkinsonian handwriting? *Human Movement Science* 10:301–314.

Plamondon, R. 1993. The generation of rapid hand movements. I. A delta lognormal law. Technical Report EPM/RT-93/94. Ecole Polytechnique de Montreal.

Plamondon, R. 1998. A kinematic theory of rapid human movements: Part III. Kinetic outcomes. *Biological Cybernetics* 78:133–145.

Plamondon, R., and Djioua, M. 2006. A multi-level representation paradigm for handwriting stroke generation. *Human Movement Science* 25:586–607.

Plamondon, R., Feng, C., and Woch, A. 2003. A kinematic theory of rapid human movement. Part IV: A formal mathematical proof and new insights. *Biological Cybernetics* 89:126–138.

Plamondon, R., and Guerfali, W. 1997. The origin of the 2/3 power law. *Proceedings of the Eighth Biennial Conference of the International Graphonomics Society*, 17–18. Genoa, Italy.

———. 1998a. The generation of handwriting with delta-lognormal synergies. *Biological Cybernetics* 78:119–132.

———. 1998b. The 2/3 power law: When and why? *Acta Psychologica* 100:85–96.

Platel, H., Lambert, J., Eustache, F., Cadet, B., Dary, M., Viader, F., and Lechevalier, B. 1993. Characteristics and evolution of writing impairment in Alzheimer's disease. *Neuropsychologia* 31:1147–1158.

Pope, P., Wing, A. M., Praamstra, P., and Miall, R. C. 2005. Force related activations in rhythmic sequence production. *Neuroimage* 27:909–918.

Potvin, A. R., Syndulko, K., Tourtellotte, W. W., Lemmon, J. A., and Potvin, J. H. 1980. Human neurologic function and the aging process. *Journal of American Geriatric Association* 29:1–9.

Poulin, G. 1999. The influence of writing fatigue on handwriting characteristics in selected populations. Part one: General considerations. *International Journal of Forensic Document Examiners* 5:193–220

Proske, U., and Gandevia, S. C. 2009. The kinesthetic senses. *Journal of Physiology* 587:4139–4146.

Provins, K. A., and Magliaro, J. 1989. Skill strength, handedness, and fatigue. *Journal of Motor Behavior* 21:113–121.

Pujol, J., Junque, C., and Vendrell, P. 1992. Reduction of the substantia nigra width and motor decline in aging and Parkinson's disease. *Archives of Neurology* 49:1119–1122.

Purtell, D. J. 1965. Effects of drugs on handwriting. *Journal of Forensic Science* 10: 335–346.

Rabin, A., and Blair, H. 1953. The effects of alcohol on handwriting. *Journal of Clinical Psychology* 9:284–287.

Raibert, M. H. 1977. Motor control and learning by the state space model. Technical report, Artificial Intelligence Laboratory, MIT (AI-TR-439).

Raja, M. 1998. Managing antipsychotic-induced acute and tardive dystonia. *Drug Safety* 19:57–72.

Rantanen, T., Masaki, K., Foley, D., Izmirlian, G., White, L., and Guralnik, J. M. 1998. Grip strength changes over 27 years in Japanese-American men. *Journal of Applied Physiology* 85:2047–2053.

Rantanen, T., Masaki, K., He, Q., Ross, G. W., Willcox, B. J., and White, L. 2011. Midlife muscle strength and human longevity up to age 100 years: A 44-year prospective study among a decedent cohort. *Age*. Published online May 4, 2011.

Rapcsak, S. Z., Arthur, S. A., Bliklen, D. A. and Rubens, A. B. 1989. Lexical agraphia in Alzheimer's disease. *Archives of Neurology* 46:65–68.

Rascol, U., Sabatini, C., Brefel, N., Fabre, N., Rai, S., Senard, J. M., Celsis, P., Viallard, G., Montastruc, J. L., and Chollet, F. 1998. Cortical motor overactivation in parkinsonian patients with l-dopa-induced peak-dose dyskinesia. *Brain* 121:527–533.

Rhodes, B. J., Bullock, D., Verwey, W. B., Averbeck, B. B., and Page, M. P. A. 2004. Learning and production of movement sequences: Behavioral, neurophysiological, and modeling perspectives. *Human Movement Science* 23:699–746.

Ricaurte, G. A., Yuan, J., Hatzidimitriou, G., Cord, B. J., and McCann, U. D. 2002. Severe dopaminergic neurotoxicity in primates after a common recreational dose regimen of MDMA ("Ecstasy"). *Science* 297:2260–2263.

Richards, J. B., Baggott, M. J., Sabol, K. E., and Seiden, L. S. 1993. A high-dose methamphetamine regimen results in long-lasting deficits on performance of a reaction-time task. *Brain Research* 627:254–260.

Rijntjes, M., Dettmers C., Buchel, C., Kiebel, S., Frackowiak, R. S. J., and Weiller, C. 1999. A blueprint for movement: Functional and anatomical representations in human motor system. *Journal of Neuroscience* 19:8043–8048.

Robichaud, J. A., Pfann, K. D., Comella, C. L., and Corcos, D. M. 2002. Effect of medication on EMG patterns in individuals with Parkinson's disease. *Movement Disorders* 17:950–960.

Rodriguez-Aranda, C. 2003. Reduced writing and reading speed and age-related changes in verbal fluency tasks. *Clinical Neuropsychologist* 17:203–215.

Roeltgen, D. P., and Heilman, K. M. 1983. Apractic agraphia in a patient with normal praxis. *Brain Language* 18:35–46.

Romero, D. H., and Stelmach, G. E. 2001. Motor function in neurodegenerative disease and aging. In *Handbook of neuropsychology*, ed. F. Boller and S. F. Capa, 163–191. Amsterdam: Elsevier Science.

Rosati, G. 2001. The prevalence of multiple sclerosis in the world: An update. *Neurology Science* 22:117–139.

Rosenbaum, D. A. 2010. *Human motor control*, 2nd ed. University Park, PA: Academic Press.

Rosenbaum, D. A., Loukopoulos, L. D., Meulenbroek, R. G. J., Vaughan, J., and Engelbrecht, S. E. 1995. Planning reaches by evaluating stored postures. *Psychological Review* 102:28–67.

Rosenbaum, D. A., Cohen, R. G., Jax, S. A., Weiss, D. J., and van der Wel, R. 2007. The problem of serial order in behavior: Lashley's legacy. *Human Movement Science* 26: 525–554.

Rosenbaum, D. A., Vaughan, J., Jorgensen, M. J., Barnes, H. J., and Stewart, E. 1993. Plans for object manipulation. In *Attention and performance XIV: Synergies in experimental psychology, artificial intelligence, and cognitive neuroscience*, ed. D. E. Meyer and S. Kornblum, 803–820. Hillside, NJ: Lawrence Erlbaum Associates.

Roser, P., Gallinat, J., Weinberg, G., Juckel, G., Gorynia, I., and Stadelmann, A. M. 2009. Psychomotor performance in relation to acute oral administration of Delta 9-tetrahydrocannabinol and standardized cannabis extract in healthy human subjects. *European Archives of Psychiatry and Clinical Neuroscience* 259:294–292.

Rothwell, J. C., Traub, M. M., and Marsden, C. D. 1979. Primary writing tremor. *Journal of Neurology Neurosurgery Psychiatry* 42:1106–1114.

Roux, F., Draper, L., Köpke, B., and Démonet, J. 2010. Who actually read Exner? Returning to the source of the frontal "writing centre" hypothesis. *Cortex* 46:1204–1210.

Rummel-Kluge, C., Komossa, K., Schwartz, S., Humger, H., Schmid, F., Kissling, W., Davis, J. M., and Leucht, S. 2010. Second generation antipsychotic drugs and extrapyramidal side effects: A systematic review and meta-analysis of head-to-head comparisons. *Schizophrenia Bulletin* May 31 [Epub ahead of print].

Sabbe, B, van Hoof, J., Hulstijn, W., andZitman, F. 1997. Depressive retardation and treatment with fluoxetine: assessment of the motor component. *Journal of Affective Disorders* 43:53–61

Sabbe, B., van Hoof, J., Hulstijn, W., and Zitman, F., 1996. Changes in fine motor retardation in depressed patients treated with fluoxetine. *Journal of Affective Disorders* 40, 149–157.

Sabbe, B., Hulstijn, W., van Hoof, J., Tuynman-Qua, H. G., and Zitman, F. 1999. Retardation in depression: Assessment by means of simple motor tasks. *Journal of Affective Disorders* 55:39–44.

Sabbe, B., Hulstijn, W., van Hoof, J., and Zitman, F. 2006. Fine motor retardation and depression. *Journal of Psychiatric Research* 30:295–306.

Sachdev, P. 1995. *Akathisia and restless legs.* Cambridge, England: Cambridge University Press.

Saling, L. L. and Phillips, J. G. 2005. Variations in the relationship between radius of curvature and velocity as a function of joint motion. *Human Movement Science* 24: 731–43.

Saling, L. L., and Phillips, J. G. 2002. Age-related changes in the kinematics of curved drawing movements: Relationship between tangential velocity and the radius of curvature. *Experimental Aging Research* 28:215–229.

Salamone, J. D., Correa, M., Farrar, A. M., Nunes, E. J., and Pardo, M. 2009. Dopamine, behavioral economics, and effort. *Frontiers in Behavioral Neuroscience* 3:13.

Salisachs, P., and Findley, L. J. 1984. Problems in the differential diagnosis of essential tremor. In *Movement disorders: Tremor,* ed. L. J. Findley and R. Capildeo, 219–224. New York: Oxford University Press.

Salthouse, T. A. 1993. Attentional blocks are not responsible for age-related slowing. *Journal of Gerontology* 48:P283–P270.

Salthouse, T. A., and Somberg, B. L. 1982. Isolating the age deficit in speeded performance. *Journal of Gerontology* 37:59–63.

Samuel, M., Ceballos, B. A., Blin, J., Uema, T., Boecker, H., Passingham, R. E., and Brooks, D. J. 1997. Evidence for lateral premotor and parietal over activity in Parkinson's disease during sequential and bimanual movements. A PET study. *Brain* 120:963–976.

Sanudo-Pena, M. C., Tsou, K., and Walker, J. M. 1999. Motor actions of cannabinoids in the basal ganglia output nuclei. *Life Science* 65:703–713.

Scarone, P., Gatignol, P., Guillaume, S., Denvil, D., Capelle, L., and Duffau, H. 2009. Agraphia after awake surgery for brain tumor: New insights into the anatomo-functional network of writing. *Surgical Neurology* 72:223–241.

Schaal, S., and Sternad, D. 2001. Origins and violations of the 2/3 power law in rhythmic three-dimensional arm movements. *Experimental Brain Research* 136:60–72.

Schmidt, R. A. 1975. A schema theory of discrete motor skill learning. *Psychological Review* 82:225–260.

Schmidt, R. A., and Lee, T. D. 1999. *Motor control and learning, a behavioral emphasis,* 3rd ed. Champaign, IL: Human Kinetics.

Schmidt, R. A., Zelanik, H. N., Hawkins, B., Frank, J. S., and Quinn, J. T. 1979. Motor output variability: A theory for the accuracy of rapid motor acts. *Psychological Review* 86:415–451.

Schretlen, D., Pearlson, G. D., Anthony, J. C., Alyward, E. H., Augustine, A. M., Davis, A., and Barta, P. 2000. Elucidating the contributions of processing speed, executive ability, and frontal lobe volume to normal age-related differences in fluid intelligence. *Journal of International Neuropsychology Society* 6:52–61.

Schrijvers, D., Hulstijn, W., and Sabbe, B. G. 2008. Psychomotor symptoms in depression: A diagnostic, pathophysiological and therapeutic tool. *Journal of Affective Disorders* 109:1–20.

Schrijvers, D., Maas, Y. J., Pier, M. P., Madani, Y., Hulstin, W., and Sabbe, B. G. 2009. Psychomotor changes in major depressive disorders during sertraline treatment. *Neuropsychobiology* 59: 34–42.

Schröter, A., Mergl, R., Bürger, K., Hampel, H., Möller, H.-J. and Hegerl, U. 2003. Kinematic analysis of handwriting movements in patients with Alzheimer's disease, mild cognitive impairment, depression and healthy subjects. *Dementia and Geriatric Cognitive Disorders* 15:132–142.

Seaman-Kelly, J., and Lindblom, B. 2006. *Scientific examination of questioned documents,* 2nd ed. Boca Raton, FL: CRC Press.

Seidler, R. D., Alberts, J. L., and Stelmach, G. E. 2002. Changes in multi-joint performance with age. *Motor Control* 6:19–31.

Seidler, R. D., Bernard, J. A., Burutolu, T. B., Fling, B. W., Gordon, M. T., Gwin, J. T., Kwak, Y., and Lipps, D. B. 2010. Motor control and aging: Links to age-related brain structural, functional, and biochemical effects. *Neuroscience and Biobehavioral Reviews* 34:721–733.

Seidler-Dobrin, R. D., He, J., and Stelmach, G. E. 1998. Coactivation to reduce variability in the elderly. *Motor Control* 2:314–330.

Seitz, R., Canavan, A. G., Yaguez, L., Herzog, H., Tellmann, L., Knorr, U., Huang, Y., and Homberg, V. 1997. Representations of graphomotor trajectories in the human parietal cortex: Evidence for controlled processing and automatic performance. *European Journal of Neuroscience* 9:278–289.

Shinohara, M., Li, S., Kang, N., Zatsiorsky, V. M., and Latach, M. L. 2003. Effects of age and gender on finger coordination in maximal contractions and submaximal force matching tasks. *Journal of Applied Physiology* 94:259–270.

Shinohara, M., Scholtz, J. P., Zatsiorsky, V. M., and Latash, M. L. 2004. Finger interaction during accurate multi-finger force production tasks in young and elderly persons. *Experimental Brain Research* 156:282–292.

Siebner, H. R., Limmer, C., Peinemann, A., Bartenstein, P., Drzezga, A., and Conrad, B. 2001. Brain correlates of fast and show handwriting in humans: A PET–performance correlation analysis. *European Journal of Neuroscience* 14:726–736.

Silventoinen, K., Magnusson, P. K., Tynelius, P., Kaprio, J., and Rasmussen, F. 2008. Heritability of body size and muscle strength in young adulthood: A study of one million Swedish men. *Genetic Epidemiology* 32:341–349.

Silveri, M. C., Corda, F., and Di Nardo, M. 2007. Central and peripheral aspects of writing disorders in Alzheimer's disease. *Journal of Clinical and Experimental Neuropsychology* 29:179–186.

Simon, S. L., Domier, C., Carnell, J., Brethen, P., Rawson, R., and Ling, W. 2000. Cognitive impairment in individuals currently using methamphetamine. *American Journal of Addiction* 9:222–231.

Sita, J., Found, B., and Rogers, D. 2002. Forensic handwriting examiners' expertise for signature comparison. *Journal of Forensic Science* 47:1117–1124.

Slavin, M. J., Phillips, J. G., and Bradshaw, J. L. 1996. Visual cues and the handwriting of older adults: A kinematic analysis. *Psychology and Aging* 11:521–526.

Slavin, M. J., Phillips, J. G., Bradshaw, J. L., Hall, K. A., and Presnell, I. 1999. Consistency of handwriting movements in dementia of Alzheimer's type: A comparison with Huntington's and Parkinson's diseases. *Journal of the International Neuropsychological Society* 5:20–25.

Slavin, M. J., Phillips, J. G., Bradshaw, J. L., Hall, K. A., Presnell, I., and Bradshaw, J. A. 1995. Kinematics of handwriting movements in dementia of the Alzheimer's type. *Alzheimer's Research* 1:123–132.

Smyth, M. M., and Silvers, G. 1987. Functions of vision in the control of handwriting. *Acta Psychologica* 3:276–315.

Sobin, C., and Sackeim, H. A. 1997. Psychomotor symptoms of depression. *American Journal of Psychiatry* 154:4–17.

Soechting, J. F., and Lacquaniti, F. 1981. Invariant characteristics of a pointing movement in man. *Journal of Neuroscience* 1:710–720.

Soland, V. L., Bhatia, K. P., Volante, M. A., and Marsden, C. D. 1996. Focal task-specific tremors. *Movement Disorders* 11:665–670.

Soliveri, P., Piacentini, S., and Girotti, F. 2005. Limb apraxia in corticobasal degeneration and progressive supranuclear palsy. *Neurology* 64:448–453.

Sosnoff, J. J., and Newell, K. M. 2006. Are age-related increases in force variability due to decrements in strength? *Experimental Brain Research* 174:86–94.

Spraker, M. B., Yu, H., Corcos, D. M., and Vaillancourt, D. E. 2007. Role of individual basal ganglia nuclei in force amplitude generation. *Journal of Neurophysiology* 98:821–834.

Stahl, S. M. 1996. *Essential psychopharmacology.* Cambridge, England: Cambridge University Press.

Stelmach, G. E., Phillips, J., DiFabio, R. P., and Teasdale, N. 1989. Age, functional postural reflexes, and voluntary sway. *Journal of Gerontology* 44:B100–B106.

Sternberg, S., Monsell, S., Knoll, R. L., and Wright, C. E. 1978. The latency and duration of rapid movement sequences: Comparisons of speech and typewriting. In *Information processing in motor control and learning,* ed. G. E. Stelmach, 117–152. New York: Academic Press.

Stetson, R. H., and McDill, J. A. 1923. Mechanism of the different types of movement. *Psychology Monographs* 32:18.

Stiles, R. N. 1976. Frequency and displacement amplitude relations for normal hand tremor. *Journal of Applied Physiology* 40:44–54.

Stinson, M. D. 1997. A validation study of the influence of alcohol on handwriting. *Journal of Forensic Science* 2:411–416.

Stringer, S. M., and Rolls, E. T. 2007. Hierarchical dynamical models of motor function. *Journal of Neurocomputing* 70:975–990.

Sugihara, G., Kaminaga, T., and Sugishita, M. 2006. Interindividual uniformity and variety of the "writing center": A functional MRI study. *Neuroimage* 32:1837–1849.

Taekema, D. G., Gussekloo, J., Maier, A. B., Westendorp, R. G. J., and Craen, A. J. M. 2010. Handgrip strength as a predictor of functional, psychological, and social health. A prospective population-based study among the oldest old. *Age and Ageing* 39:331–337.

Tarver, J. A. 1988. *Micrographia in the handwriting of Parkinson's disease patients.* Presented at the meeting of the American Society of Questioned Document Examiners (Aurora, CO).

Teulings, H. L. 1996. Handwriting movement control. In *Handbook of perception and action. Vol. 2: Motor skills,* ed. S. W. Keele and H. Heuer, 561–613. London: Academic Press.

Teulings, H. L., Contreras-Vidal, J. L. Stelmach, G. E., and Adler, C. H. 2002. Handwriting size adaptation under distorted visual feedback in Parkinson's disease patients, elderly and younger controls. *Journal of Neurology Neurosurgery and Psychiatry* 72:315–324.

Teulings, H.L., Contreras-Vidal, J.L., Stelmach, G.E., and Adler, C.H. (1997). Coordination of fingers, wrist, and arm in Parkinsonian handwriting. *Experimental Neurology*, 146, 159–170.

Teulings, H. L., and Maarse, F. J. 1984. Digital recording and processing of handwriting movements. *Human Movement Science* 3:193–217.

Teulings, H. L. and Schoemaker, L.R.B. 1993. Invariant properties between stroke features in handwriting. *Acta Psychologica* 82: 69–88.

Teulings, H. L., and Stelmach, G. E. 1991. Force amplitude and force duration in parkinsonian handwriting. In *Tutorials in motor neurosciences,* ed. J. Requin and G. E. Stelmach, 149–160. Dordrecht, The Netherlands: Kluwer.

Teulings, H. L., and Thomassen, A. J. W. M. 1979. Computer-aided analysis of handwriting movements. *Visible Language* 13:219–231.

Thach, W. T. 1998. A role for the cerebellum in learning movement coordination. *Neurobiology of Learning and Memory* 70:177–188.

Thaler, D. S. 2002. Design for an aging brain. *Neurobiology of Aging* 23:13–15.

Thobois, S., Ballanger, B., Baraduc, P., Le Bars, D., Lavenne, F., Broussolle, E., Desmurget, M., et al. 2007. Functional anatomy of motor urgency. *Neuroimage* 37:243–252.

Thomassen, A. J. W. M., and Shoemaker, L. R. B. 1986. Between-letter context effects in handwriting trajectories. In *Graphonomics: Contemporary research in handwriting,* ed. H. S. R. Kao, G. P. Van Galen, and R. Hoosain, 253–263. Amsterdam: Elsevier.

Thomassen, A. J. W. M., and Teulings, H. L. 1985. Time size and shape in handwriting: Exploring spatio-temporal relationships at different levels. In *Time, mind, and behavior,* ed J. A. Michon and J. L. Jackson, 253–263. Berlin: Springer.

Thomassen, A. J. W. M., and van Galen, G. P. 1992. Handwriting as a motor task: Experimentation, modeling and simulation. In *Approaches to the study of motor control and learning,* ed. J. J. Summers, 113–144. Amsterdam: North–Holland.

Tiplady, B., Baird, R., Lutche, H., Drummond, G., and Wright, P. 2005. Effects of ethanol on kinesthetic perception. *Journal of Psychopharmacology* 9:627–632.

Tiraboschi, P., Salmon, D. P., Hansen, L. A., Hofstetter, R. C., Thal, L. J., and Corey-Bloom, J. 2006. What best differentiates Lewy body from Alzheimer's disease in early-stage dementia? *Brain* 129:729–735.

Toomey, R., Lyons, M. J., Eisen, S. A., Xian, H., Chantarujakipong, S., Seidman, L. J., Faraone, S. V., and Tsuang, M. T. 2003. A twin study of the neuropsychological consequences of stimulant abuse. *Archives of General Psychiatry* 6:303–310.

Tracy, B. L., Maluf, K. A., Stephenson, J. L., Hunter, S. K., and Enoka, R. M. 2005. Variability of motor unit discharge and force fluctuations across a range of muscle forces in older adults. *Muscle and Nerve* 32:533–540.

Tripp, C. A., Fluchiger, F. A., and Weinberg, G. H. 1959. Effects of alcohol on the graphomotor performance of normals and chronic alcoholics. *Perceptual and Motor Skills* 9:227–236.

Tucha, O., Aschenbrenner, S., Eichhammer, P., Putzhammer, A., Sartor, H., Klein, H. E., and Lange, K. W. 2002. The impact of tricyclic antidepressants and selective serotonin reuptake inhibitors on handwriting movements of patients with depression. *Psychopharmacology (Berl)* 159:211–215.

Tucha, O., Mecklinger, L., Thome, J., Reiter, A., Alders, G. L., Sartor, H., Naumann, M., and Lange, K. W. 2006. Kinematic analysis of dopaminergic effects on skilled handwriting movements in Parkinson's disease. *Journal of Neural Transmission* 113:609–623.

Turner, R. S., and Anderson, M. E. 1997. Pallidal discharge related to the kinematics of reaching movements in two dimensions. *Journal of Neurophysiology* 77:1051–1074.

Turner, R. S., and Desmurget, M. 2010. Basal ganglia contributions to motor control: A vigorous tutor. *Current Opinion in Neurology* 20:1–13.

Turner, R. S., Desmurget, M., Grethe, J., Crutcher, M. D., and Grafton, S. T. 2003. Motor subcircuits mediating the control of movement extent and speed. *Journal of Neurophysiology* 90:3958–3966.

Tytell, P. 1998. Pen pressure as an identifying characteristic of signatures: Verification from the computer. *Journal of American Society of Questioned Document Examiners* 1:21–31.

Umegaki, H., Roth, G. S., and Ingram, D. K. 2008. Aging of the striatum: Mechanisms and interventions. *Age* 30:251–261.

Uno, Y., Kawato, M., and Suzuki, R. 1989. Formation and control of optimal trajectory in human multi-joint arm movement. *Biological Cybernetics* 61:89–101.

UNODC 2008. http://www.unodc.org/documents/about-unodc/AR08_WEB.pdf

Vaillancourt, D. E., Larsson, L., and Newell, K. M. 2003. Effects of aging on force variability, single motor unit discharge patterns and the structure of 10, 20 and 40 Hz EMG activity. *Neurobiology of Aging* 24:25–35.

Valenstein, E., and Heilman, K. M. 1979. Apraxic agraphia with neglect-induced paragraphia. *Archives of Neurology* 36:506–508.

van Dyck, C. H., Avery, R. A., MacAvoy, M. G., Marek, K. L., Quinlan, D. M., Baldwin, R. M., Seibyl, J. P., Innis, R. B., and Arnsten, A. F. 2009. Striatal dopamine transporters correlate with simple reaction time in elderly subjects. *Neurobiology of Aging* 29:1237–1246.

van Dyck, C. H., Seibyl, J. P., Malison, R. T., Laruelle, M., Zoghbi, S. S., Baldwin, R. M., and Innis, R. B. 2002. Age-related decline in dopamine transporters: Analysis of striatal subregions, nonlinear effects, and hemispheric asymmetries. *American Journal of Geriatric Psychiatry* 10:36–43.

van Galen, G., and van Gemmert, A. 1996. Kinematic and dynamic features of simulating another person's handwriting. *Journal of Forensic Document Examination* 9:1–25.

van Galen, G. P. 1991. Handwriting: Issues for psychomotor theory. *Human Movement Science* 10:165–191.

van Galen, G. P., Meulenbroek, R. G. J., and Hylkema, H. 1986. On the simultaneous processing of words, letters and strokes in handwriting: Evidence for mixed linear and parallel models. In *Graphonomics: Contemporary research in handwriting,* ed. H. S. R. Kao, G. P. van Galen, and R. Hoosain, 5–20. Amsterdam: North–Holland.

van Galen, G. P., and Teulings, H.-L. 1983. The independent monitoring of form and scale factors in handwriting. *Acta Psychologica* 54:9–22.

van Galen, G. P., and Weber, J. F. 1998. On-line size control in handwriting demonstrates the continuous nature of motor programs. *Acta Psychologica* 100:195–216.

van Gemmert, A., Adler, C., and Stelmach, G. 2003. Parkinson's disease patients undershoot target size in handwriting and similar tasks. *Journal of Neurology, Neurosurgery, & Psychiatry* 74:1502–1508.

van Gemmert, A., van Galen, G., Hardy, H., and Thomassen, A. 1996. Dynamical features of disguised handwriting. Presented at the Fifth European Conference for Police and Government Handwriting Experts, The Hague, the Netherlands.

van Gemmert, A. W. A., Teulings, H. L., Contreras-Vidal, J. L., and Stelmach, G. E. 1999. Parkinson's disease and the control of size and speed of handwriting. *Neuropsychologia* 37:685–694.

van Gemmert, A. W. A., Teulings, H. L., and Stelmach, G. E. 2001. Parkinsonian patients reduce their stroke size with increased processing demands. *Brain and Cognition* 47:504–512.

van Hoof, J. J., Hulstijn, W., van Mier, H., and Pagen, M. 1993. Figure drawing and psychomotor retardation: Preliminary report. *Journal of Affective Disorders* 29:263–266.

van Nes, F. L. 1971. Errors in the motor programme for handwriting. *IPO Annual Progress Report* 6:61–63.

Verrel, J., Lovden, M., and Lindenberger, U. 2010. Normal aging reduced motor synergies in manual pointing. *Neurobiology of Aging* August 17 [Epub ahead of print].

Vinter, A., and Mounoud, P. 1991. Isochrony and accuracy of drawing movements in children; effects of age and context. In *Development of graphic skills*, ed. J. Wann, A. M. Wing, and N. Sõvik, 113–114. London: Academic Press.

Viviani, P. 1986. Do units of motor action really exist? In *Generation and modulation of action patterns*, ed. H. Heuer and C. Fromm, 201–216. Berlin: Springer–Verlag.

Viviani, P., Burkhard, P. R., Chiuve, S. C., dell'Acqua, C. C., and Vindras, P. 2009. Velocity control in Parkinson's disease: A quantitative analysis of isochrony in scribbling movements. *Experimental Brain Research* 194:259–283.

Viviani, P., and Flash, T. 1995. Minimum-jerk, two-thirds power law, and isochrony: Converging approaches to movement planning. *Journal of Experimental Psychology: Human Perception and Performance* 21:32–53.

Viviani, P., and McCollum, G. 1983. The relation between linear extent and velocity in drawing movements. *Neuroscience* 10:211–228.

Viviani, P., and Schneider, R. 1991. A developmental study of the relationship between geometry and kinematics in drawing movements. *Journal of Experimental Psychology* 17:198–218.

Viviani, P., and Terzuolo, C. 1980. Space–time invariance in learned motor patterns. In *Tutorials in motor behavior*, ed. G. A. Stelmach and J. Requin, 525–533. Amsterdam: North–Holland.

———. 1982. Trajectory determines movement dynamics. *Neuroscience* 7:431–437.

Volkow, N. D., Chang, L., Wang, G.-J., Fowler, J. S., Leonido-Yee, M., Franceschi, D., Sedler, M. J., et al. 2001. Association of dopamine transporter reduction with psychomotor impairment in methamphetamine abusers. *American Journal of Psychiatry* 158:377–382.

Volkow, N. D., Ding, Y. S., Fowler, J. S., Wang, G. J., Logan, J., Gatley, S. J., Hitzemann, R., Smith, G., Fields, S. D., and Gur, R. 1996. Dopamine transporters decrease with age. *Journal of Nuclear Medicine* 37:554–559.

Volkow, N. D., Gur, R. C., Wang, G. J., Fowler, J. S., Moberg, P. J., Ding, Y. S., Hitzemann, R., Smith, G., and Logan, J. 1998. Association between decline in brain dopamine activity with age and cognitive and motor impairment in healthy individuals. *American Journal of Psychiatry* 155:344–349.

Volkow, N. D., Wang, G-J., Franceschi, D., Fowler, J. S., Thanos, P. K., Maynard, L., Gatley, S. J., et al. 2006. Low doses of alcohol substantially decreases glucose metabolism in the human brain. *Neuroimage* 29:295–301.

Wade, P., Gresty, M. A., and Findley, L. J. 1982. A normative study of postural tremor of the hand. *Archives of Neurology* 39:358–362.

Walker, F. O. 2007. Huntington's disease. *Lancet* 369:220.

Walton, J. 1997. Handwriting changes due to aging and Parkinson's syndrome. *Forensic Science International* 88:197–214.

Wang, Y., Chan, G. L., Holden, J. E., Dobko, T., Mak, E., Schulzer, M., Huser, et al. 1998. Age-dependent decline of dopamine D-1 receptors in human brain: A PET study. *Synapse* 30:56–61.

Wann, J., Nimmo-Smith, I., and Wing, A. M. 1988. Relation between velocity and curvature in movement: Equivalence and divergence between a power law and a minimum-model. *Journal of Experimental Psychology: Human Perception and Performance* 14:622–637.

Weigner, A. W., and Wierzbicka, M. M. 1992. Kinematic models and elbow flexion movements: Quantitative analysis. *Experimental Brain Research* 88:665–673.

Wellingham-Jones, P. 1991. Characteristics of handwriting of subjects with multiple sclerosis. *Perception of Motor Skills* 73:867–879.

Wendt, G. 2000. Statistical observations of disguised signatures. *Journal of American Society of Questioned Document Examiners* 3:19–27.

Werner, P., Rosenblum, S., Bar-On, G., Heinik, J., and Korczyn, A. 2006. Handwriting process variables discriminating mild Alzheimer's disease and mild cognitive impairment. *Journal of Gerontology B: Psychological Sciences and Social Sciences* 61:228–236.

Wilkinson, R. T., and Allison, S. 1989. Ages and simple reaction time: Decade differences for 5,325 subjects. *Journal of Gerontology* 44:P29–P35.

Willard, V. 1997. Parkinson's disease and graphic disturbances. *Journal of Forensic Document Examination* 10:1–40.

Wing, A. M. 1980. The height of handwriting. *Acta Psychologica* 46: 141–151.

Wing, A. M. 2000. Motor Control: Mechanisms of motor equivalence in handwriting. *Current Biology*. 10: R245–R248

Woch, A., Plamondon, R., and O'Reilly, C. 2011. Kinematic characteristics of bidirectional delta-lognormal primitives in young and older subjects. *Human Movement Science* 30:1–17.

Wolfson, L., Whipple, R., Derby, C. A., Amerman, P., Murphy, T., Tobin, J. N., and Nashner, L. 1992. A dynamic posturography study of balance in healthy elderly. *Neurology* 42:2069–2075.

Wong, D. F., Broussolle, E. P., Wand, G., Villemagne, V., Dannals, R. F., Links, J. M., Zacur, H. A., et al. 1998. In vivo measurement of dopamine receptors in human brain by positron emission tomography. *Annals of New York Academy of Sciences* 515:203–214.

Wright, C. E. 1993. Evaluating the special role of time in the control of handwriting. *Acta Psychologica* 82:5–52.

Yan, J. H., Rountree, S., Massman, P., Doody, R. S., and Li, H. 2008. Alzheimer's disease and mild cognitive impairment deteriorate fine movement control. *Journal of Psychiatric Research* 42:1203–1212.

Yassa, R., and Jeste, D. V. 1992. Gender differences in tardive dyskinesia: A critical review of the literature. *Schizophrenia Bulletin* 18:701–715.

Zaki, N. N., and Ibraheim, M. A. 1983. Effect of alcohol and *Cannabis sativa* consumption on handwriting. *Neurobehavioral Toxicology Teratology* 5:225–227.

Index

B

C

T